普通高等教育"十三五"规划教材

高分子化学实验
Polymer Chemistry Experiments (Bilingual)
（双语版）

张晓云　曲剑波　编著

中国石化出版社
HTTP://WWW.SINOPEC-PRESS.COM

内容提要

本书内容为中英文对照，全书共分为三个部分，第一章介绍了高分子化学实验基础知识，内容包括实验室安全常识、常用的聚合反应装置、常用的原料精制方法、聚合物提纯方法及表征手段。第二章为高分子化学实验部分，共引入验证性、探究性和实用性等各类实验项目21个。第三章是文献检索部分，主要介绍了SciFinder、ISI Web of Knowledge、常用外文数据库网站及重要的高分子学术期刊等文献检索资源。

本书可作为高分子化学、材料化学、应用化学、化学、化工及相关专业的实验教材，也可供从事高分子材料研究和应用的工程技术人员参考。

图书在版编目（CIP）数据

高分子化学实验：双语版 / 张晓云，曲剑波编著.
—北京：中国石化出版社，2020.11
ISBN 978–7–5114–6017–2

Ⅰ.①高… Ⅱ.①张…②曲… Ⅲ.①高分子化学–化学实验–高等学校–教材–汉、英 Ⅳ.① O63-33

中国版本图书馆CIP数据核字（2020）第222312号

未经本社书面授权，本书任何部分不得被复制、抄袭，或者以任何形式或任何方式传播。版权所有，侵权必究。

中国石化出版社出版发行
地址：北京市东城区安定门外大街58号
邮编：100011 电话：(010)57512500
发行部电话：(010)57512575
http://www.sinopec-press.com
E-mail：press@sinopec.com
北京科信印刷有限公司印刷
全国各地新华书店经销

*

710×1000毫米16开本 12印张 172千字
2020年11月第1版 2020年11月第1次印刷
定价：39.00元

前　言

高分子科学的形成和发展始于20世纪20年代，目前高分子合成技术发展迅猛，合成的各种功能性高分子材料已在国防建设、工农业生产以及生物医药等领域有着举足轻重的地位，这对高校培养具有创新精神的应用型高分子化学人才提出了更高的要求。高分子化学实验作为高分子化学的配套课程，可以将抽象的基础理论与直观的实验结果联系起来，具有较强实践性和探索性，有利于培养学生的实验技能和创新思维，对于理论教学具有重要的推动作用，也是新工科背景下培养高素质创新型人才不可或缺的重要环节。

目前国内外已出版了不少《高分子化学实验》教材，对于高分子相关专业的学生掌握基本实验技术和技能，具有重要的应用价值和指导意义。随着高分子科学的不断发展和"新工科"背景下人才培养模式的多样化，《高分子化学实验》教材有必要适应新的发展形势。为此，本书在前人基础上，引入探究性和实用性实验，主要目的是注重培养学生独立分析问题和解决问题的能力，提高学生的创新思维和综合素质，满足当前社会对于高素质的应用型和创新型人才的需求，这也是高分子化学实验改革的一个重要方向。

本书的另外一个特色是双语性，书中的实验内容都是中英文对照，有助于培养学生英语描述相关实验的能力，提高学生专业英语水平，为学生阅读相关英文文献奠定基础。另外本书还介绍了 Sci Finder 等文献检索资源的使用，有助于培养学生掌握文献检索的能力。

本教材共分为三个部分。第一章介绍了高分子化学实验相关基础知识，目的是使学生掌握基本的常识和实验操作；第二章是实验部分，引入验证性、探究性和实用性等各类实验项目 21 个，包括实验目的、实验原理、实验器材、实验步骤、附注和思考题。第三章是文献检索部分，主要介绍了 SciFinder、ISI Web of Knowledge、常用外文期刊网站检索资源。

本书的编写参考了国内外相关教材和文献资料，并引用了一些重要的公式和方法，在此向各位专家前辈和兄弟院校的同行表示诚挚谢意。另外，本书的插图由曲嘉睿同学提供，对此深表谢意。

由于编者水平有限，加之时间仓促，书中难免出现错误和疏漏情况，恳请各位专家、同行以及广大读者提出宝贵意见。

目　录

第一章　高分子化学实验基础 001

1.1　高分子化学实验室安全常识 003
1.1.1　人身防护 003
1.1.2　危险化学品 003
1.1.3　废弃物的处理 005
1.1.4　消防安全 005

1.2　常用的聚合反应装置 007
1.2.1　玻璃仪器 007
1.2.2　聚合反应装置 007

1.3　常用的原料精制方法 011
1.3.1　洗涤 011
1.3.2　减压蒸馏 013
1.3.3　快速分离柱 015
1.3.4　重结晶 015

1.4　聚合物常用提纯方法 017

1.5　聚合物的表征 019

第二章　实验部分 021

实验一　甲基丙烯酸甲酯铸板聚合 023
实验二　甲基丙烯酸甲酯本体聚合反应速率的定性观测 029
实验三　悬浮聚合法制备有机玻璃模塑粉 035
实验四　有机玻璃的解聚 039
实验五　苯乙烯－甲基丙烯酸甲酯共聚反应竞聚率的测定 043

实验六　不饱和聚酯树脂和玻璃纤维增强塑料的制备 053
实验七　丙烯酰胺水溶液聚合 063
实验八　超高吸水性材料——低交联聚丙烯酸钠的合成 067
实验九　聚醋酸乙烯酯胶乳的制备 071
实验十　聚氨酯泡沫塑料的制备 075
实验十一　苯乙烯的阳离子聚合 083
实验十二　苯乙烯的可逆加成-断裂链转移（RAFT）聚合 089
实验十三　聚合物表征——黏度 097
实验十四　缩聚法制备尼龙66 103
实验十五　甲基丙烯酸甲酯本体聚合动力学 109
实验十六　尼龙的界面聚合 115
实验十七　自修复柔性透明导电薄膜的制备 119
实验十八　冠醚辅助苯乙烯的无皂乳液聚合 127
实验十九　超高分子量聚丙烯酰胺的合成 133
实验二十　表面活性剂反胶团溶胀法制备超大孔聚苯乙烯微球 139
实验二十一　溶液聚合法制备温敏性耐高温体膨颗粒 145

第三章　文献检索 153

3.1　SciFinder"数据库" 155
　　3.1.1　SciFinder"搜索" 159
　　3.1.2　SciFinder"定位" 163
　　3.1.3　SciFinder"浏览" 165
3.2　ISI Web of Knowledge 平台 167
3.3　常用的外文期刊网站 169

附录Ⅰ　高分子相关的主要学术期刊 174
附录Ⅱ　常见聚合物的中英文对照及缩写 176
附录Ⅲ　一些常见的高分子溶剂和沉淀剂 179
附录Ⅳ　常见引发剂的提纯方法 181
参考文献 182

CONTENTS

Chapter 1　Experimental Basis of Polymer Chemistry 001

  1.1　Common safe sense of Polymer Chemistry Laboratory 002
      1.1.1　Personal safety protection 002
      1.1.2　Hazardous chemicals 002
      1.1.3　Waste disposal 004
      1.1.4　Fire safety 004
  1.2　Common polymerization devices 006
      1.2.1　Glasswares 006
      1.2.2　Polymerization devices 006
  1.3　Common raw material refining methods 010
      1.3.1　Washing 010
      1.3.2　Vacuum distillation 012
      1.3.3　Fast separation column 014
      1.3.4　Recrystallization 014
  1.4　Common polymer purification methods 016
  1.5　Characterization of polymer 018

Chapter 2　Experiment Part 021

  Experiment 1　Casting polymerization of methyl methacrylate 022
  Experiment 2　Investigate the rate of bulk polymerization of methyl
                    methacrylate qualitatively 028
  Experiment 3　Suspension polymerization of PMMA for molding powder 034
  Experiment 4　Depolymerization of Poly(methyl methacrylate) 038

Experiment 5	Determination of reactivity ratio of styrene-methyl methacrylate copolymer	042
Experiment 6	Synthesis of unsaturated polyester resin and glass fiber reinforced plastics	052
Experiment 7	Solution polymerization of acrylamide	062
Experiment 8	Synthesis of super absorbent material —— low cross-linking sodium polyacrylate	066
Experiment 9	Synthesis of polyvinyl acetate latex	070
Experiment 10	The preparation of polyurethane foam plastics	074
Experiment 11	Cationic polymerization of styrene	082
Experiment 12	Reversible addition-fragmentation chain transfer (RAFT) polymerization of styrene	088
Experiment 13	Polymer Characterization—Viscosimetry	096
Experiment 14	Condensation polymerization: preparation of nylon 66	102
Experiment 15	Kinetics of bulk polymerization of methyl methacrylate	108
Experiment 16	Syntgesis of Nylon by interfacial polymerization	114
Experiment 17	Fabrication of self-healing flexible transparent conductive film	118
Experiment 18	Soap-Free Emulsion Polymerization of Styrene with the assistance of crown ethers	126
Experiment 19	Synthesis of polyacrylamide with ultrahigh molecular weight	132
Experiment 20	Synthesis of gigaporous poly(styrene-divinyl benzene) microspheres by surfactant reverse micelles swelling method	138
Experiment 21	Synthesis of thermosensitive high-temperature resistant superabsorbent via solution polymerization	144

Chapter 3　Literature Search ... 153

3.1　SciFinder ... 154
　　3.1.1　SciFinder 'Explore' ... 158
　　3.1.2　SciFinder 'Locate' ... 162
　　3.1.3　SciFinder 'Browse' ... 164
3.2　ISI Web of Knowledge ... 166
3.3　Common websites of foreign database ... 168

第一章
Chapter 1

高分子化学实验基础
Experimental Basis of Polymer Chemistry

Polymer chemistry, as the fifth chemistry after inorganic chemistry, organic chemistry, analytical chemistry and physical chemistry, is a fundamental course for students majored in chemical engineering, chemistry and materials. The matching course of polymer chemistry experiment can not only deepen students' understanding of basic theoretical knowledge, exercise students' basic experimental skills, but also improve students' innovation ability by introducing some exploratory experiments. It plays an important role in promoting theoretical teaching and is an indispensable part in cultivating high-quality innovative talents under the background of "emerging engineering".

Polymer chemistry is developed on the basis of organic chemistry, but it has its own characteristics. Before introducing polymer chemistry experiments, it is necessary to briefly introduce some basic knowledge involved in polymer chemistry experiments.

1.1 Common safe sense of Polymer Chemistry Laboratory

1.1.1 Personal safety protection

Chemistry experiments have certain risks. When conducting chemical experiments, it is necessary to wear necessary protective equipment, including experimental clothes, safety glasses and gloves, which must be fully worn. It is not permitted to wear shorts and sandals (or sports shoes with mesh) in the laboratory. You must remove gloves and wash hands before leaving the laboratory. It is not allowed to wear gloves when entering the instrument analysis room for sample analysis, and it is not allowed to touch the door handle with gloved hands to prevent cross contamination.

In addition, the students should understand the distribution of and operation of safety facilities equipped in the laboratory, such as first aid kit, eye washer, shower equipment, fire extinguisher and oxygen concentration meter, etc.

1.1.2 Hazardous chemicals

You must look up instructions of the chemicals involved in the experiment, fully understand the physical and chemical characteristics of the chemicals, and use the chemicals strictly in accordance with the instructions. Especially for toxic chemical reagents, the laboratory should have necessary protective measures to ensure its safe storage and use.

The storage of chemical reagents should follow the principle of classification: solid reagents, liquid reagents, organic solvents, highly toxic drugs, etc. All of them have standard storage instructions. For

高分子化学作为继无机化学、有机化学、分析化学和物理化学之后的第五大化学，是化工、化学、材料类专业的基础课程。与之配套的高分子化学实验，不仅能够加深学生对基础理论知识的理解，锻炼学生的基本实验技能，更能通过引入一些探究性实验来提高学生的创新能力，对于理论教学具有重要的推动作用，是"新工科"背景下培养高素质创新型人才不可或缺的环节。

高分子化学是在有机化学的基础上发展起来的，但又有其自身的特点，在介绍高分子化学专业实验之前，有必要对高分子化学实验中涉及的一些基础知识进行简要介绍。

1.1 高分子化学实验室安全常识

1.1.1 人身防护

化学实验具有一定的危险性，进行化学实验时必须穿戴必要的防护器具，包括实验服、护目镜、手套等。在实验室内不允许穿短裤和凉鞋（或带网孔的运动鞋），离开实验室时必须摘除手套洗手后方可离开。进入仪器分析室进行样品分析时不允许戴手套，不得用戴手套的手接触门把手，防止交叉污染。另外，还需要了解实验室配备的安全设施的分布和使用，如急救箱、洗眼器、冲淋设备、灭火器和氧气浓度测定仪等。

1.1.2 危险化学品

对于实验涉及的化学药品，使用前要详细查阅使用说明，充分了解化学药品的物理和化学特性，严格按照说明使用化学药品。特别是对于有毒化学试剂，实验室要具备必要的防护措施，保证其安全存放和使用。

化学试剂的存放遵循分门别类原则：固体试剂、液体试剂、有机溶剂、剧毒药品等都有标准的存放说明。例如，烯类单体和自由基引发剂性质活泼，应

example, alkene monomers and free radical initiators with active chemical characters must be stored in refrigerators. Photosensitive initiators should be kept away from light. Strong oxidants and reducers should not be put together, halogenated hydrocarbons and alkali metals should be stored separately, flammable and explosive solvents should be placed in explosion-proof iron cabinets. In addition, special personnel should manage highly toxic drugs, and there are "five pairs" principles: double storage, double lock, double account, double picking and double use.

The toxic chemicals involved in polymer chemistry experiments mainly include aromatic compounds (benzene, nitrobenzene, chlorobenzene, aniline, phenylhydrazine, polycyclic aromatic hydrocarbons, etc.), halogen compounds (chloroform, carbon tetrachloride, dichloroethane, sulfoxide chloride, bromoethane, etc.), nitrogen compounds (acetonitrile, cyanide, nitroso compounds, etc.), highly carcinogenic substances (benzene, ethylene oxide, chloromethane ether, toluene-2, 4-diisocyanate, etc.).

1.1.3 Waste disposal

Randomly discarding the waste liquid and waste residue produced in the experiment process not only brings heavy burden to the urban sewage treatment and garbage treatment system, but also seriously pollutes the environment, endangers the health and even pollutes the domestic water. It is very necessary to establish the environmental awareness of "one for all and all for one". The following regulations should be complied with: ① Do not pour waste liquids into the sewer, and there should be special wast tanks for the collection of wast liquids; ② Put the waste acids and alkali liquids separately; ③ Collect organic waste liquid containing halogen separately from that without halogen; ④ Do not pour waste liquid containing heavy metal into the water system; ⑤ Do not discard sharp substances, such as glass chips, needles, blades, etc., to gengral bins, and special recycled containers are necessary; ⑥ Ordinary solid waste such as chromatography supports and inorganic salt used for drying can be dumped as garbage besides the waste residues of heavy metals or other toxic substances; ⑦ The remaining active metals (sodium, magnesium, calcium hydride, sodium hydride, etc.) must be quenched before further treatment to prevent fire.

1.1.4 Fire safety

Laboratory managers should teach students to use the fire extinguishers (dry powder extinguishers, CO_2 foam extinguishers, etc.) and the correct use of fire blanket. In case of fire, keep calm, correctly judge the fire situation, evacuate in time and call 119 for fire alarm if there is major danger; when the fire situation is controllable, make full use of the fire fighting equipment in the laboratory to quench the fire, prevent the fire from spreading and minimize the loss.

存储在冰箱中；光敏引发剂要避光保存；强氧化剂和强还原剂不能放在一起；卤代烃和碱金属要分开存储；易燃易爆溶剂应放置在防爆铁皮柜中；剧毒药品要有专人管理，实行"五双"制度，即双人保管、双锁、双账、双人领取、双人使用。

高分子化学实验中涉及的有毒化学品主要有芳烃类化合物（苯、硝基苯、氯苯、苯胺、苯肼、多环芳烃等），含卤素化合物（氯仿、四氯化碳、二氯乙烷、氯化亚砜、溴乙烷等），含氮化合物（乙腈、氰化物、亚硝基化合物等），高度致癌物（苯、环氧乙烷、氯甲醚、甲苯-2,4-二异氰酸酯等）。

1.1.3 废弃物的处理

随意丢弃实验过程中产生的废液和废渣不仅给城市污水处理和垃圾处理系统带来沉重负担，而且严重污染环境，危害健康，甚至污染生活用水，要树立"我为人人，人人为我"的环保意识。必须遵守下列原则：①废液严禁倒入下水道，要有专门的废液桶回收；②废酸、废碱液要分开回收；③含卤素的有机废液要与不含卤素的分开收集；④含有重金属的废液不得倒入下水系统；⑤尖锐物质，如玻璃残片、针头、刀片等要用专门的容器回收，不得丢于普通垃圾袋中；⑥普通固体废渣，如层析填料和干燥用的无机盐可作为垃圾倒掉，含有重金属或其他有毒物质的废渣例外；⑦反应剩余的活泼金属（钠、镁、氢化钙、氢化钠等）必须淬灭后再进一步处理，防止发生火灾。

1.1.4 消防安全

实验室管理人员要教会学生灭火器（干粉灭火器、CO_2泡沫灭火器等）和灭火毯的正确使用方法。一旦失火要保持镇静，正确判断火势情况，有重大危险时要及时撤离，并拨打火警电话119；火势不大可以控制时，要充分利用实验室的消防器材进行灭火，防止火势扩大蔓延，将损失降到最低。

1.2 Common polymerization devices

1.2.1 Glasswares

Polymer chemical reactions are generally performed in grinded glass instruments, which are divided into inner grinded glass instruments and outer grinded glass instruments. At present, the grinded glass instruments have been standardized, including $10^{\#}$, $12^{\#}$, $14^{\#}$, $19^{\#}$, $24^{\#}$, $29^{\#}$, $34^{\#}$, $40^{\#}$, $45^{\#}$, and $50^{\#}$. The unit of grinded port diameter is mm. Common ground reaction bottles include single neck round bottom flask, two-neck round bottom flask, three-neck round bottom flask, four-neck round bottom flask, eggplant flask, pear shaped flask, reaction tube, reaction bottle, reaction kettle, etc. (Figure 1.1).

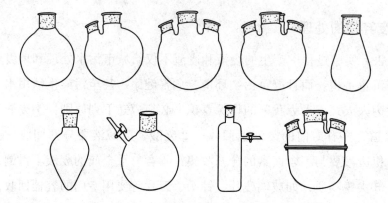

Figure 1.1　Common grinded reaction bottles

1.2.2 Polymerization devices

Most of the polymerization devices are the same as the general chemical experimental devices. Based on the reaction bottle, some auxiliary devices, such as mechanical stirrer, agitator, condenser, titration funnel, thermometer, inflation tube, etc., can be built into a polymerization device. Figure 1.2 shows six common polymerization devices. For the reaction without deaeration, a three-neck round bottom bottle equipped with one condensing tube and thermometer is feasible (Fig. 1.2a), and it can also be equipped with a titration funnel to control the reaction rate (Fig. 1.2b). For the polymerization system with deaeration requirements, nitrogen charging (Fig. 1.2c) or combined deaeration by vacuumizing and nitrogen charging (Fig. 1.2d) can be used. For solution polymerization system with low viscosity of reaction solution, magnetic stirrer can be used instead of top mechanical stirring (Fig. 1.2 e, f). In order to obtain high monomer conversion, some polycondensation reactions need to be carried out at high temperature and high vacuum in the later stage to remove small molecules. A group of reduced pressure

1.2 常用的聚合反应装置

1.2.1 玻璃仪器

高分子化学反应一般都是在磨口玻璃仪器中完成，磨口玻璃仪器分为内磨口和外磨口，目前已经标准化，有 $10^\#$、$12^\#$、$14^\#$、$19^\#$、$24^\#$、$29^\#$、$34^\#$、$40^\#$、$45^\#$ 和 $50^\#$ 等磨口，磨口直径的单位是 mm。常见的磨口反应瓶有单口圆底烧瓶、双口圆底烧瓶、三口圆底烧瓶、四口圆底烧瓶、茄型烧瓶、梨形烧瓶、反应瓶、反应管、分体式反应釜等（图1.1）。

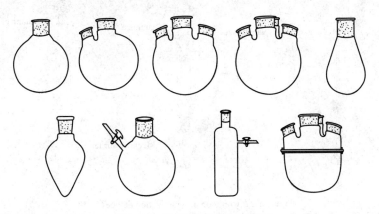

图 1.1 常见的磨口反应瓶

1.2.2 聚合反应装置

大多数高分子反应装置与普通化学实验装置相同，在反应瓶的基础上加上一些辅助器件，如搅拌器、搅拌桨、冷凝管、滴液漏斗、温度计、充气管等就可以搭建成聚合反应装置，6种常用的聚合反应装置如图1.2所示。对于不需除氧的反应，配冷凝管和温度计即可（图1.2 a），还可配上滴液漏斗控制反应速率（图1.2 b）。对于有除氧要求的聚合体系，可以采用充氮气除氧（图1.2 c）或抽真空—充氮气联合除氧（图1.2 d），对于反应液黏度不大的溶液聚合体系，可以用磁力搅拌器代替顶端机械搅拌（图1.2 e，f）。对于某些缩聚反应，反应后期需要在高温高真空条件下进行，以除去小分子物质达到较高转化

polymerization device can be built under this condition (Fig. 1.3).

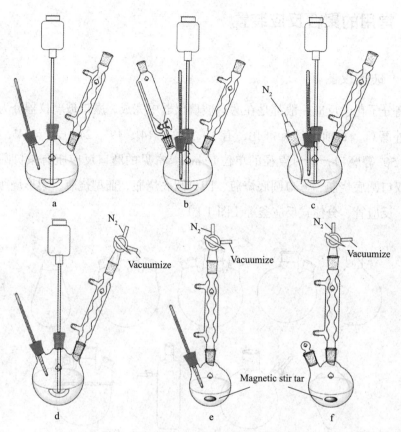

Figure 1.2 Common polymerization devices

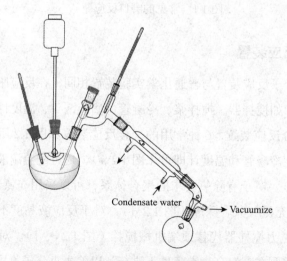

Figure 1.3 Reduced pressure polymerization device

率，这时可以搭建一组减压聚合反应装置，如图 1.3 所示。

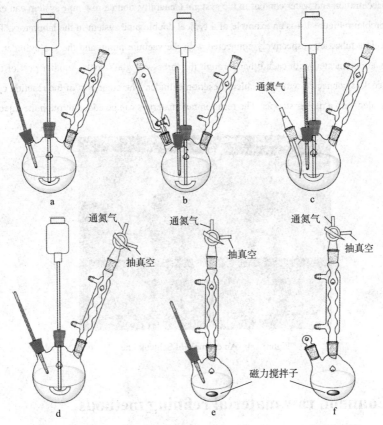

图 1.2　常用聚合反应装置

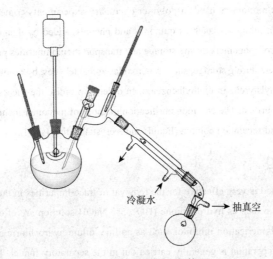

图 1.3　减压聚合反应装置

Ring opening polymerization and controlled/living polymerization have strict requirements on oxygen concentration and water content in the system. Generally, double row pipe system can effectively solve the problem. Figure 1.4 is an example of a typical double pipe system in the laboratory. The upper and lower glass tubes are respectively connected with the vacuum pump and the gas cylinder. The two glass tubes are connected with each other through the three-way glass cock. Another port of the three-way glass cock is connected with the Schlenk reaction cylinder. The other port of the reaction cylinder is generally sealed with a rubber stopper. The reaction raw materials can be added through the injector.

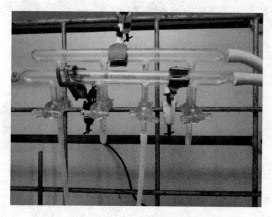

Figure 1.4 An example of Schlenk line

1.3 Common raw material refining methods

The impurities in monomers used in polymer chemistry experiments come from the following sources: polymerization inhibitors, such as quinones and phenols, added by manufacturers in order to prevent polymerization of monomers during storage and transportation; impurities produced by oxidation or reduction of monomers during storage, such as peroxides and aldehydes; by products introduced in the production process (ethylstyrene in diethylbenzene). In order to promise the successful polymerization, the monomer must be purified. The common purification methods of monomer include washing, vacuum distillation (liquid), rapid separation column (liquid) and recrystallization (solid).

1.3.1 Washing

The washing method is very effective for the removal of trace impurities in the monomer. For acid polymerization inhibitor such as hydroquinone (HQ), 5% NaOH solution are effective to remove the inhibitor. For basic polymerization inhibitor such as aniline, dilute hydrochloric acid can be used for washing. The washing operation is generally carried out in the separatory funnel. The water content of

高分子化学反应中的开环聚合、活性聚合对体系内氧气和水的含量有严格要求，通常采用双排管系统可以有效解决这个问题。图1.4是实验室中一个典型的双排管系统示例图。上下两根玻璃管分别与真空泵和气瓶相连，两根玻璃管之间通过三通活塞相连，三通活塞的另外一个接口与Schlenk反应瓶相连，反应瓶的另外一个接口一般用翻口橡皮塞密封，反应原料可通过注射器加入。

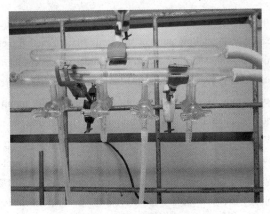

图1.4 双排管装置示例图

1.3 常用的原料精制方法

高分子化学实验所用单体中杂质的主要来源有：生产厂家为了防止单体在贮存和运输过程中发生聚合反应而加入的阻聚剂，如醌类和酚类阻聚剂；单体在贮存过程中发生氧化或还原反应而产生的杂质，如过氧化物和醛类等；生产过程中引入的副产物（二乙烯苯中的乙基苯乙烯）。为了保证聚合反应顺利进行必须对单体进行提纯，常用的单体提纯方法有洗涤、减压蒸馏（液体）、快速分离柱（液体）和重结晶（固体）等。

1.3.1 洗涤

洗涤法对于除去单体中的微量杂质非常有效，对于对苯二酚（HQ）等酸性阻聚剂可以用5%的NaOH溶液洗涤，对于苯胺等碱性阻聚剂可采用稀盐酸洗涤除去，洗涤操作一般在分液漏斗中进行。单体经过洗涤后含水量会增加，需要加入干燥剂（无水氯化钙、无水硫酸镁、无水硫酸钠等）除去单体中的水

the monomer will increase after washing. It is necessary to add desiccant (anhydrous calcium chloride, anhydrous magnesium sulfate, anhydrous sodium sulfate, etc.) to remove the residual water in the monomer. If the water content is strictly required, you can choose calcium hydride as desiccant. After filtering out the desiccant, the obtained monomer can be used for polymerization by further vacuum distillation.

1.3.2 Vacuum distillation

In the experiment of polymer chemistry, the boiling point of some liquid alkene monomers exceeds 100 ℃. For example, the boiling point of styrene is 145 ℃, methyl methacrylate is 100.5 ℃, and N, N-dimethylaminoethyl methacrylate is 186 ℃. In order to prevent thermal polymerization of monomers during distillation process, vacuum distillation are necessary to refine such monomers. There are generally two kinds of vacuum pumping equipment: water pump (1~100 kPa) can be selected when the vacuum requirement is low, and oil pump (< 1 kPa) must be used when the vacuum requirement is high. When using the oil pump, it is necessary to install a drying tower to absorb water vapor, acid gas and low boiling point solvent before the oil pump, so as to prolong the service life of the oil pump. During vacuum distillation, the liquid volume should not exceed 1/2 of the volume of the distillation bottle. Zeolite, stirrer or capillary can prevent the liquid from sudden boiling. In addition, the heating rate and vacuum degree should be adjusted. A simple vacuum distillation unit is shown in Figure 1.5.

Take the purification of styrene as an example (1) add 300 mL of styrene into a 500 mL separating funnel, wash styrene with 50 mL of 5% NaOH solution for several times until the water layer changes from red to colorless, and styrene presents light yellow; (2) wash styrene to neutral with 50 mL of deionized water for several times, and then add anhydrous sodium sulfate to dry overnight; (3) filter thedried styrene and conduct vacuum distillation directly, collect the fraction and put it into the refrigerator at –20 ℃ for reserve.

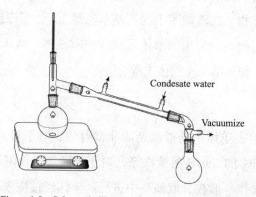

Figure 1.5　Schematic illustration of vacuum distillation unit

分，如果对含水量要求严格可加入氢化钙，将得到的单体进一步减压蒸馏即可用于聚合。

1.3.2 减压蒸馏

在高分子化学实验中，有些液体烯类单体沸点超过了 100℃，如苯乙烯的沸点为 145℃，甲基丙烯酸甲酯的沸点为 100.5℃，甲基丙烯酸 N，N- 二甲氨基乙酯的沸点为 186℃。为了防止蒸馏过程中单体的热聚合，提纯这类单体时可采用减压蒸馏方式。抽真空的设备一般有两种：真空度要求较低时可选用水泵（1~100 kPa），真空度要求较高时必须使用油泵（<1 kPa）。使用油泵时需要在油泵前安装分别吸收水蒸气、酸性气体和低沸点溶剂的干燥塔，延长油泵的使用寿命。减压蒸馏时液体不可超过蒸馏瓶体积的 1/2，容器内放置沸石、搅拌子或插入毛细管可以防止液体暴沸，另外还需要注意调节升温速度和真空度。简单的减压蒸馏装置如图 1.5 所示。

以苯乙烯的精制过程为例

（1）在 500 mL 分液漏斗中加入 300 mL 苯乙烯，用 50 mL 的 5% NaOH 溶液多次洗涤苯乙烯，直至水层由红色变为无色，苯乙烯呈淡黄色。

（2）用 50 mL 去离子水将苯乙烯洗涤至中性，然后加入无水硫酸钠干燥过夜；

（3）将干燥后的苯乙烯过滤后直接进行减压蒸馏，收集馏分后放入 -20℃ 冰箱保存备用。注意：精制后的苯乙烯不宜长久放置，因为去除阻聚剂后容易自聚。

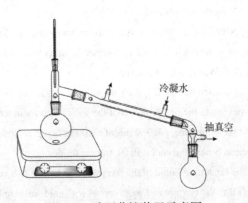

图 1.5 减压蒸馏装置示意图

1.3.3 Fast separation column

Hydroquinone (HQ), hydroquinone monomethyl ether (MEHQ) and p-tert-butylphenol (PTBP) are common polymerization inhibitors, which can be easily removed by using basic aluminum oxide column. The specific steps are as follows: take a glass pipette commonly used in the laboratory, plug the absorbent cotton on the tip, pack 10 cm high alkaline aluminum oxide particles (200-300 mesh) into it to get a fast separation column, and push the monomer through the column with an rubber suction bulb.

1.3.4 Recrystallization

Recrystallization is an effective refining method for solid monomers, initiators and catalysts. The principle of recrystallization is introduced in detail in organic chemistry, which will not be discussed here. The choice of solvent and emperature control are crucial for recrystallization. Firstly, solvents must be inert and do not react with solids. Secondly, the solubility of the solvent to the target substance is very small at room temperature, and the solubility increases gradually at elevated temperature. Finally, higher solubility of impurity in the solvent is preferable, which can ensure the impurity remains in the mother liquor when the purified substance crystallizes out. Recrystallization solvent can be a single solvent or a mixed solvent. Common recrystallization solvents include ethanol, acetone, ether, dichloromethane, n-hexane, ethyl acetate, etc. The recrystallization temperature is directly related to the solubility, and the increase of temperature favors to the dissolution of the target product. However, the dissolution temperature cannot be higher than 50 ℃ for the monomer and initiator in order to prevent thermal polymerization or initiator decomposition. The volume of solvent is usually 5%~10% higher than the actual saturated solvent volume at the corresponding dissolving temperature. After dissolving the target solid matter, the solution should be filtered while hot. In addition, it is necessary to preheat the filtering device in the oven in advance to reduce the loss in the filtration process. The crystal size can be controlled by the cooling rate. The crystal particles obtained by rapid cooling are small, and large crystal particles can be obtained by slow cooling.

Take the refining of N-isopropylacrylamide (NIPAM) as an example

(1) Dissolve 50 g of NIPAM in 30 mL of mixed solvent composed of 15 mL acetone (solvent) and 15 mL n-hexane (precipitant) in a beaker (200 mL) immersed in water bath at 40 ℃. Preheat the Buchner funnel and the suction flask (250 mL) at the same time.

(2) After complete dissolution of NIPAM, filter the solution while hot using the preheated Buchner funnel. Add 35 mL of n-hexane to the filtrate and cool the solution to room temperature before put it in a freezer for one night. At the same time, put 360 mL of washing liquid composed of 30 mL acetone and 330 mL n-hexane into a clean beaker and cool it in the freezer for one night.

(3) Decant most of the remaining liquid in the recrystallized beaker to a conical flask and break the NIPAM crystal carefully. Filter the mixture and wash crystal with cold washing liquid at the same time.

1.3.3 快速分离柱

对苯二酚（HQ）、对苯二酚单甲醚（MEHQ）、对叔丁基苯酚（PTBP）是经常使用的阻聚剂，这类阻聚剂采用碱性三氧化二铝柱可以方便地除去。具体步骤：取一根实验室常用的玻璃吸管，在尖端塞上脱脂棉，装入 10 cm 高碱性三氧化二铝颗粒（200~300 目）得到快速分离柱，用吸耳球将单体推送过柱即可用于聚合。

1.3.4 重结晶

对于固体单体、引发剂和催化剂，重结晶除杂是一种很有效的方法。重结晶的原理在有机化学中有详细介绍，这里不再赘述。重结晶的关键是溶剂的选择和温度的控制。首先，溶剂必须是惰性的，不与固体物质发生反应。其次，溶剂在常温下对被提纯物质的溶解度很小，升高温度时溶解度逐渐增大。最后，溶剂对杂质的溶解度越大越好，可以保证被提纯物质结晶析出时杂质仍然留在母液中。重结晶溶剂可以是单一溶剂，也可以是混合溶剂。常用的重结晶溶剂有乙醇、丙酮、乙醚、二氯甲烷、正己烷、乙酸乙酯等。重结晶温度与溶解度直接相关，温度升高有利于被提纯物的溶解，但对于高分子单体和引发剂，溶解温度不能超过 50℃，以防止热聚合和引发剂分解。在溶解温度下，溶剂量比实际饱和溶剂量多 5%~10% 即可。固体物质溶解后要趁热过滤，为了降低过滤过程中的损失，需要将过滤装置提前放在烘箱中预热。晶体的大小可通过降温速度控制，急剧降温得到的晶体颗粒小，缓慢降温得到的晶体颗粒大。

以 N- 异丙基丙烯酰胺（NIPAM）的精制为例。

（1）称取 50g NIPAM 放入 200 mL 烧杯中，加入丙酮（溶剂）和正己烷（沉淀剂）各 15 mL，在 40℃水浴中溶解；同时将布氏漏斗和抽滤瓶（250mL）预热。

（2）待 NIPAM 完全溶解后，趁热抽滤；将滤液倒入干净烧杯，向滤液中加入 35 mL 正己烷，待溶液冷却至室温后放入冰柜冷藏一晚；同时，分别取 30 mL 丙酮和 330 mL 正己烷混合（洗涤液），放入冰柜冷藏一晚；

（3）将析出 NIPAM 晶体后的剩余液体小心倒出，再将晶体弄碎，将晶体真空抽滤，抽滤的同时用洗涤液淋洗晶体；

(4) Dry the refined NIPAM in a vacuum drying oven and store it in the freezer for further use.

1.4 Common polymer purification methods

In order to characterize and further apply the polymer synthesized, it is generally necessary to purify the polymer from the reaction system after polymerization. The purification method of polymer is different from monomer, initiator and other small molecules due to the characteristic of high molecular weight of polymer. The common separation and purification methods for polymer include filtration washing, centrifugal washing, dissolution precipitation, rotary evaporation, solvent extraction, dialysis, etc. The above methods can be used alone or in combination.

Polymers synthesized with precipitation polymerization, suspension polymerization, interfacial polymerization and dispersion polymerization etc., are in the form of precipitation in the reaction system, which can be separated by means of filtration washing or centrifugal washing. Repeated centrifugal washing and dialysis are effective methods for purification of nanoparticles synthesized by emulsion polymerization or self-assembly. Dialysis method is to use the property that small molecules can pass through the dialysis bag in the solution, while high polymer is trapped in the dialysis bag. By changing the dialysate, all small molecules can be removed finally. When choosing dialysis bag, we should pay attention to the material and molecular weight of dialysis bag. To ensure that not only the dialysis bag is insoluble in the dialysate, but also its cut-off molecular weight is lower than the molecular weight of the target polymer.

If the polymer synthesized is soluble in the reaction system, the rotary evaporator can be used to concentrate the reaction system first. Then, the concentrated polymer solution is added dropwise to precipitant under stirring. The precipitated polymer can be collected by filtration or centrifugation. The volume of precipitant is generally 4~10 times that of polymer solution.

For crosslinked polymer particles, solvent extraction can effectively remove unreacted monomers and oligomers. Solvent extraction is usually carried out in Soxhlet extractor (Fig. 1.6). The glasswares from top to bottom in Figure 1.6 are condenser tube, extractor and distillation flask, respectively. Briefly, the sample is wrapped with filter paper or cloth and put into the extractor, and 2/3 volume of solvent is filled in the distillation flask. After heating, the extraction solvent vapor rises from the steam pipe will be condensed into liquid by the condenser tube. The condensate drops back to the extractor and dissolve the soluble components in the polymer particles. When the liquid level is higher than the highest point of the siphon, the solvent in the extractor will all be siphoned into the distillation flask, and the next round of dissolution extraction process starts. After extraction for a period of time, all soluble components can be extracted into the distillation flask, and the polymer particles in the extractor can be purified.

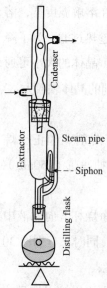

Figure 1.6　Schematic of soxhlet extractor

（4）将晶体室温下真空干燥后，放入冰箱冷藏备用。

1.4 聚合物常用提纯方法

聚合反应完成后，为了对聚合物进行表征和进一步应用，一般都需要把聚合物从聚合体系中分离出来。聚合物属于高分子，它的分离纯化与单体、引发剂等小分子不同。常用的分离纯化方法有过滤洗涤法、离心洗涤法、溶解沉淀法、旋转蒸发法、溶剂抽提法、透析法等，上述方法可以组合使用。

对于以沉淀形式存在于反应体系中的产品（如沉淀聚合、悬浮聚合、界面缩聚、分散聚合等），可采用过滤洗涤法或离心洗涤法进行分离提纯，对于乳液聚合得到的纳米粒子，采用多次离心洗涤或透析可以分离提纯。透析法是利用小分子物质在溶液中可通过透析袋，而高分子聚合物被截留在透析袋的性质，通过更换透析液，最终使小分子全部被去除。选择透析袋时要注意透析袋的材质和截留分子量，要保证透析袋不被透析液溶解，聚合物分子量要大于透析袋的截留分子量。

如果聚合物溶解于反应体系，可先采用旋转蒸发仪浓缩反应体系，接着将浓缩反应液以滴加方式加入不良溶剂，使聚合物沉淀后再过滤或离心分离。沉淀剂的体积一般为聚合物溶液体积的 4~10 倍。对于交联的聚合物颗粒，采用溶剂抽提可以有效去除未反应的单体和低聚物。溶剂抽提一般在索氏抽提器（图 1.6）中进行。图 1.6 中自上而下依次是冷凝管、提取器和蒸馏瓶。样品用滤纸或滤布包好放入抽提器中，蒸馏瓶中装入 2/3 溶剂，加热后溶剂蒸气由蒸气导管上升至冷凝管中被冷却，滴落回流至抽提器中，聚合物中的可溶性组分被溶解。当液面高于虹吸管最高点时，抽提器溶剂被全部虹吸到蒸馏瓶中，开始下一轮的溶解提取过程。抽提一段时间后就可以将所有可溶性组分抽提到蒸馏瓶中，抽提器内的聚合物得到纯化。

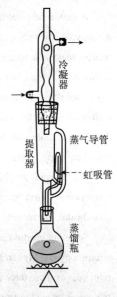

图 1.6 索氏抽提器

1.5 Characterization of polymer

It is necessary to characterize the properties of purified polymer to verify the accuracy of the experimental results. The characterization contents of polymer generally include chemical structure composition, molecular weight and molecular weight distribution, thermal stability and morphology, etc. Fourier transform infrared spectrum (FT-IR), raman spectroscopy (RS), X-ray photoelectron spectroscopy (XPS) and nuclear magnetic resonance (NMR) are usually used to characterize the chemical structure and composition of the polymer.

There are many methods to determine the molecular weight of polymer, including membrane osmotic pressure method, gas phase osmotic method, boiling point rise method, freezing point reduction method, light scattering method, ultracentrifugation sedimentation method, viscosity method, matrix assisted laser desorption time of flight mass spectrometer (MALDI-TOF-MS), etc. Among them, gel permeation chromatography (GPC) and viscosity method are most often used molecular weight measurement methods in the laboratory. Both methods need to use the same polymer with known molecular weight as a benchmark to get absolute molecular weight. Gel permeation chromatography (GPC) is also called size exclusion chromatography. It relies on the principle of physical separation. It separates the polymer with different molecular weights according to retention time: the smaller the retention time is, the larger the molecular weight is, and vice versa.

It should be notable that traditional gel permeation chromatography has only one concentration detector. It can only detect the hydrodynamic volume of polymer in the mobile phase, calculate the relative molecular weight, and ignore the difference of polymer structure (such as branching, association, conformation, etc.), which makes the experimental value far away from the theoretical value. Modern gel permeation chromatography (e.g. VISCOTEK) is usually equipped with multi detection system (differential detector, viscosity detector, light scattering detector and UV detector). It can not only give the absolute molecular weight of samples, but also provide information about viscosity, aggregation and branching of polymers.

The thermal stability analysis of polymer can be characterized by TGA. Under programmed temperature control, the relationship between polymer mass and temperature (or time) can be recorded to give the thermal stability, decomposition temperature and residue content of the material.

Scanning electron microscopy (SEM), transmission electron microscopy (TEM) and atomic force microscopy (AFM) can be used to observe the morphology of polymer nanospheres and micelles. The principles of these three instruments are different, and the sample preparation methods are also different. In general, special instrument operators are required for sample preparation and analysis.

1.5 聚合物的表征

聚合物得到分离提纯后，需要对其进行性质表征，以验证实验结果的准确性。聚合物的表征内容一般有：化学结构组成、分子量大小及分子量分布、热稳定性以形貌表征等。聚合物的化学结构及组成表征可以采用红外光谱（FT-IR）、拉曼光谱（RS）、X射线光电子能谱（XPS）、核磁共振（NMR）等。

聚合物分子量的测定方法有很多，包括膜渗透压法、气相渗透法、沸点升高法、冰点降低法、光散射法、超速离心沉降法、黏度法、基质辅助激光解吸—飞行时间质谱仪（MALDI-TOF-MS）等。实验室常用的测定聚合物的分子量方法是凝胶渗透色谱法和黏度法。这两种方法都需要用已知分子量的同种聚合物作为基准才能得到绝对分子量。凝胶渗透色谱又叫尺寸排阻色谱，它依靠单纯的物理分离原理，将不同分子量的高分子按照保留时间分开，保留时间越小的高分子分子量越大，保留时间越大的高分子分子量越小。需要注意的是，传统的凝胶渗透色谱只有一个浓度检测器，它只能检测到聚合物在流动相中的流体力学体积，得到的是相对分子量，忽视了聚合物结构的不同（如支化、缔合、构象等），从而使实验值远远偏离理论值。现代凝胶渗透色谱仪（如VISCOTEK）通常配备多检测系统，示差检测器、黏度检测器、光散射检测器和紫外检测器，不仅可以给出样品的绝对分子量，还能提供聚合物特性黏度、聚集和支化方面的信息。

聚合物的热稳定性分析可以热重分析仪（TGA）表征，在程序控温下，测量高分子的质量随温度（或时间）的变化关系，从而可以给出材料的热稳定性、分解温度以及残留物含量。

对于聚合物微纳米球、胶束等聚集体结构，可以采用扫描电镜（SEM）、透射电镜（TEM）以及原子力显微镜（AFM）观察它们的形貌。这三种仪器原理不同，制样方式也不一样，一般需要专门的仪器操作人员进行制样分析。

第二章
Chapter 2

实验部分
Experiment Part

Experiment 1 Casting polymerization of methyl methacrylate

1. The purposes

(1) Be familiar with the principles and methods to produce plexiglass using bulk polymerization.

(2) Know the effects of automatic acceleration on bulk polymerization.

2. Experimental principle

Cast polymerization is a form of bulk polymerization, and we can get plastic sheet directly by this method. Due to poly (methyl methacrylate) sheet erspex's high transparency, hypobarism, good insulation, explosion-proof and shock-proof, etc, it has a very wide range of applications, such as windshield, sunroof, instrument guards, medical guide light pipes, optical lenses and the like.

Remarkable features of bulk polymerization are that the high viscosity and poor heat-transfer ability of polymerization system, and there is automatic acceleration phenomenon in a certain stage during the reaction. At this time, if the heat of reaction can not be exhausted it can broaden the distribution of molecular weight, reduce mechanical strength of material. Sometimes the implosion could let product become useless.

The process of casting polymerization is generally divided into three stages. The first stage is pre-polymerization. System has low viscosity and poor heat-transfer ability, and the reaction can be carried out under the condition of large container, higher temperature and agitation in the early stage of pre-polymerization. This is to get a higher polymerization rate during the initial polymerization and viscosity of the system, thus reduce the possibility of leakage of plasma after filling the mold and volume contraction. The second reaction stage occurs when it get close to automatically acceleration point. In order to avoid automatic acceleration, make the polymerization temperature decrease, and so does the reaction rate. The third stage is post-polymerization. In this stage conversion rate of monomers is in a high level and reaction could be completed even at higher temperatures.

3. Instruments and materials

Instruments: 7 cm×6 cm thick glass plate (2 pieces); tray balance; beaker (500 mL and 100 mL); 150 mL separating funnel; electronic balance; round-bottom flask; condenser; stoppers; thermometer; thermometer casing; formwork units; water bath; distillation head; vacuum tube.; vacuum pump; Erlenmeyer flask (100 mL); oven.

实验一　甲基丙烯酸甲酯铸板聚合

一、实验目的

（1）熟悉用本体聚合法制造有机玻璃板的原理和方法。
（2）了解自动加速效应对本体聚合反应的影响。

二、实验原理

铸板聚合为本体聚合法的一种形式，用这种方法可以直接得到塑料板材。聚甲基丙烯酸甲酯板材即有机玻璃板，因其具有透明度高、密度小、绝缘性能好、防爆、防震等优点，在各领域有着广泛的应用，如飞机风挡、天窗、仪表防护罩，医用导光管、光学镜片等。

本体聚合的一个显著特点是聚合体系黏度大、传热差，且在反应进行到某一阶段时会出现自动加速现象。这时若不能及时排除反应热，轻则使分子量分布变宽，降低材料的机械强度，重则引起"爆聚"而使产品报废。

铸板聚合反应一般分 3 个阶段进行。第一阶段为预聚合，聚合初期体系黏度小、传热好，反应可以在较大容器中、较高温度和有搅拌的条件下进行。这样既可以在聚合初期获得较高的聚合速率，又可以适当增加体系的黏度，从而减小灌模后漏浆的可能性及成型时的体积收缩。第二阶段是当反应进行到接近自动加速点时，降低聚合温度，减缓反应速度，以避免自动加速现象的发生。第三阶段为后期聚合，这时单体的转化率已相当高，为使聚合反应完全，这一阶段可以在较高的温度下进行。

三、仪器和试剂

仪器：7cm×6cm 厚玻璃板（2 块）；托盘天平；烧杯（500mL 和 100mL）；150 mL 分液漏斗；电子天平；圆底烧瓶；冷凝管；橡皮塞；温度计；温度计套管；铁架台；水浴锅；蒸馏头；减压接应管；真空泵；锥形瓶（100 mL）；烘箱。

Materials: 5%NaOH; anhydrous Na_2SO_4; methyl methacrylate; benzoyl peroxide; silicone oil; PVA glue.

Others: White paper; kraft paper.

4. Experimental procedure

Part 1: Refining of raw materials

(1) Purification of methyl methacrylate.

Methyl methacrylate is a kind of liquid that is clear and colorless, and it's boiling point is between 100.3-100.6 ℃, melting point is -48.2℃, specific gravity of the pure is 0.936 (20/4 ℃), refractive index is 1.4136 (20 ℃). It is slightly soluble in water and absolute soluble in alcohol, ether or other organic solvent.

In commercial methyl methacrylate, there are generally polymerization inhibitors, commonly used is hydroquinone, which can be washed away with alkali solution. The method of purification is: Add 50 mL methyl methacrylate into 150 mL separating funnel and wash several times with 10 mL 5% NaOH aqueous solution until colorless, then wash with distilled-water (each time 10 ~ 15 mL) until neutral. After separating the water layer, transfer the oil layer to the brown reagent bottle and add about 5% anhydrous Na_2SO_4, shaking thoroughly, and left to dry for more than 24h. Collect 50 ℃ and 16.5 kPa fractions by vacuum distillation, and the boiling points of methyl methacrylate at different pressure are shown in Table 2.1.

Table 2.1 Boiling point of methyl methacrylate at different pressures

Temperature/℃	20	30	40	50	60	70	80	90
Pressure/kPa	4.67	7.07	10.80	16.53	25.2	37.2	52.93	72.93

Alkaline polymerization inhibitor contained in monomer can also be removed using strong base anion exchange resin. Make use of the method of refractometric, bromination chromatography to test the purity of refined methyl methacrylate monomer.

(2) Purification of benzoyl peroxide

Solubility of benzoyl peroxide (BPO) in several solvents is shown in Table 2.2. Common method of purification is to choose a solvent which has larger solubility to BPO, and make BPO dissolved in the solvent at room temperature then saturated (no heating!). Then, put the saturated BPO solution into another bad solven for BPO and refined BPO will crystallize. Because acetone and ether have the ability to induce decomposition of BPO, so they are not suitable for recrystallization.

试剂：5% NaOH；无水 Na_2O_4；甲基丙烯酸甲酯；过氧化苯甲酰；硅油；聚乙烯醇胶水。

其他：白纸；牛皮纸。

四、实验内容

1. 原料的精制

（1）甲基丙烯酸甲酯的精制

甲基丙烯酸甲酯是无色透明的液体，沸点为 100.3~100.6 ℃，熔点为 –48.2 ℃，纯品相对密度为 0.936（20 ℃ /4 ℃），n_D^{20}=1.4136，微溶于水，易溶于乙醇和乙醚等有机溶剂。

在商品甲基丙烯酸甲酯中，一般都含有阻聚剂，常用的是对苯二酚，可用碱溶液洗去。具体步骤：在 150 mL 分液漏斗中加入 50 mL 甲基丙烯酸甲酯，用 10 mL 5% 的 NaOH 水溶液洗涤数次直至无色，然后用蒸馏水洗（每次 10~15 mL）至中性。分尽水层后转移至棕色试剂瓶中，加入单体量5% 左右的无水 Na_2SO_4，充分摇动，放置干燥 24 h 以上。减压蒸馏收集 50 ℃、16.5 kPa 的馏分，甲基丙烯酸甲酯的沸点和压力的关系见表 2.1。

表 2.1 甲基丙烯酸甲酯在不同压力下的沸点

温度/℃	20	30	40	50	60	70	80	90
压力/kPa	4.67	7.07	10.80	16.53	25.2	37.2	52.93	72.93

单体中所含的碱性阻聚剂也可以用强碱性阴离子交换树脂将其除去。精制后的单体甲基丙烯酸甲酯可用测定其折射率法、溴化法或气相色谱法检验其纯度。

（2）过氧化苯甲酰的精制

过氧化苯甲酰（BPO）在几种溶剂中的溶解度见表 2.2。精制的常用方法是选用一种溶解度较大的溶剂，在室温下使 BPO 在其中溶解并饱和（不要加热！），然后将 BPO 在另一溶解度较小的溶剂中结晶出来。由于丙酮和乙醚对 BPO 有诱导分解作用不宜作为它的重结晶溶剂。

Table 2.2 Solubility of benzoyl peroxide in several solvents

Solvent	Solubility/(g/100 mL)	Solvent	Solubility/(g/100 mL)
Petroleum ether	0.5	Toluene	11.0
Methanol	1.0	Benzene	14.6
Ethanol	1.5	Acetone	16.4
		Chloroform	31.6

Example: 4 g of BPO is dissolved in 17 mL chloroform at room temperature, then filter and remove the insoluble substance. The filtrate is poured into 50 mL methanol and the needle white crystal is filtered by a buchner funnel and washing with a small amount of methanol for three times. If necessary, repeat the crystallization process again. Dry the crystal naturally, then put the crystal in a vacuum oven to dry at room temperature. Finally, product is stored in a brown bottle for use.

Part 2: Process of casting polymerization

(1) Molding: Take 2 washed and dried glass plates while one side of plate is coated with a thin layer of silicone oil (mold release agent). There are two rubbers to regulate glass plates to the same level (regulating shim thickness depending on the thickness of desired product) between them. Brush PVA glue on white paper evenly, which is 20 mm wide and whose length is equal to the circumference of the glass. Around the two pieces of glass, paste them to make regular intervals. Firstly, paste three edges, leaving aside for filling spout. Carefully expel bubbles between paper and glasses, and use the same method to paste a layer of white paper or the kraft paper after drying. After completely dried, take out the rubber tube to get the mold. Paste it with the paper pulp after filling it from the other side.

(2) Slurrying: Put 35 g distilled methyl methacrylate and 35 mg recrystallized benzoyl peroxide into a conical flask (250 mL). The mouth of the bottle is plugged with a rubber stopper with a bag of cellophane and a vent pipe. Heat the flask in a water bath at 80-90 ℃ while gently shaking intermittently. Process of pre-polymerization takes about 30-60 min. When the viscosity of the reaction system reaches two times of the viscosity of glycerol, cool the conical flask to room temperature to slow down the reaction[1].

(3) Filling and forming: The mentioned prepolymer is poured into the mold cavity slowly and expel bubbles at the same time. After filling, connect the paper. Put the mold into the oven at 40 ℃ and keep reaction going on for 20 h. When the system is solidified, heat it to 80 ℃ and keep for 2 h, then to 100 ℃

---

[1] It should be quickly cooled immediately if it is found that viscosity is too large. After cooling, add a small amount of monomer to dilute the solution if it is not easy to fill mold.

表 2.2　过氧化苯甲酰在几种溶剂中的溶解度

溶剂	溶解度 / (g/100 mL)	溶剂	溶解度 / (g/100 mL)
石油醚	0.5	甲苯	11.0
甲醇	1.0	苯	14.6
乙醇	1.5	丙酮	16.4
		氯仿	31.6

实例：将 4 g BPO 在室温下溶于 17 mL 氯仿中，过滤，除去其中不溶物。滤液倒入 50 mL 的甲醇中，将白色针状结晶用布氏漏斗抽滤，用少量甲醇洗涤结晶 3 次，必要时可重复结晶一次。晶体自然晾干后再放入真空烘箱中，在室温下真空干燥。产品放在棕色瓶中保存于干燥器内备用。

2. 铸板聚合过程

（1）制模

取两块玻璃板洗净、烘干，在板的一面涂上一薄层硅油（脱模剂）。在玻璃板间用两段橡胶管垫平（根据所需制品的厚度调节垫片的厚度）。取宽为 20 mm、长度等于玻璃片周长的白纸条，均匀地刷上聚乙烯醇胶水后，沿两块玻璃片四周将其粘住使间隔固定。先粘贴三边，留一边作灌料口。仔细赶去纸和玻璃板间的气泡，吹干或烘干后用同样的方法再糊上一层白纸、一层牛皮纸。待完全烘干后，取出橡皮管即成模子。灌完浆料后再用纸将另一边粘贴住。

（2）制浆

在 250 mL 锥形瓶中加入新蒸的甲基丙烯酸甲酯 35 g 及经重结晶的过氧化苯甲酰 35mg，瓶口用一包有玻璃纸并装有透气管的橡皮塞塞好，充分混合后将锥形瓶置于 80~90℃ 的水浴上加热，并间断地加以轻轻摇动。预聚合反应大约需要 30~60 min。当反应体系黏度相当于两倍甘油的黏度时，将锥形瓶放到冷水中冷却至室温，使反应减缓进行❶。

（3）灌浆与成型

将上述预聚物通过注料口缓缓灌入模腔内，注意让气泡排净。待灌满后，将开口处的纸条接好，压平。将模子置于 40℃ 的烘箱中，在该温度下继续聚合 20h。待体系固化后升温至 80℃ 继而 100℃ 各保温 2 h，然后打开烘箱，令其自

---

❶ 若发现体系黏度过大，则应立即迅速冷却。冷却后若不易灌模，可加少量单体加以稀释。

and for 2 h. Open the oven to cool to room temperature ❶.

(4) Mold unloading: Loosen the paper around four sides and pry the glass open with a knife carefully. Then wash and blow-dry the finished product.

Questions

(1) What are the advantages of the method to produce plexiglass by bulk polymerization in the industry?

(2) Try to explain the reasons of self-accelerating effect. How to control the process of self-accelerating?

(3) Anlyze the factors that affect the product quality.

Experiment 2 Investigate the rate of bulk polymerization of methyl methacrylate qualitatively

1. The purposes

(1) Understand the basic principles of bulk polymerization.

(2) Investigate the main factors influencing rate of methyl methacrylate bulk polymerization qualitatively.

2. Experimental principle

Bulk polymerization of the vinyl monomer occurs in the absence of any other solvent, and dispersant with a small amount of an initiator or light, heat, radiation, ect. to initiate polymerization. The feature of bulk polymerization is that it has high polymerization rate and molecular weight, purity of the product and can be shaped directly. However, the heat of reaction releases toughly during polymerization. In addition, in some cases the raw is difficult to dissolve in monomers or there are side reactions in high viscosity, which limit the application of this method.

Polymerization of vinyl monomers initiated by the initiator is generally consistent with the following kinetic equation (1).

$$R = k[M][I]^{1/2} \tag{1}$$

Where, R is the total rate of polymerization [mol/(L·s)], k is rate constant, [M] and [I] are concentration of monomer and initiator (mol/L), respectively.

---

❶ In the second stage, i.e. low temperature polymerization stage, the reaction temperature should be strictly controlled and kept below 45 ℃. Only when all products lose liquidity can the temperature rise.

然冷却至室温❶。

（4）脱模

揭去四周的纸条，用小刀小心地撬开玻璃板，将制品取出洗净、吹干。

思考题

（1）工业上采用本体聚合的方法来制备有机玻璃有什么优点？

（2）解释自动加速效应产生的原因。如何控制自动加速过程？

（3）分析影响制品质量的因素。

实验二　甲基丙烯酸甲酯本体聚合反应速率的定性观测

一、实验目的

（1）了解本体聚合反应的基本原理。

（2）定性观测甲基丙烯酸甲酯本体聚合反应速率的主要因素。

二、实验原理

烯类单体的本体聚合是在不加任何溶剂和其他分散剂的条件下，用少量引发剂或用光、热或辐射等引发的聚合反应。本体聚合的特点是可以获得高的聚合速率和高的分子量，产物的纯度高并能直接成型。但由于聚合过程中释放出来的热传递困难，以及在某些情况下聚合物难溶于单体或高黏度下有副反应发生等情况而限制了这一方法的应用。

由引发剂引发的烯类单体的聚合一般都符合下面的动力学方程（1）：

$$R=k[M][I]^{1/2} \tag{1}$$

式中，R 为聚合总速率，$mol/(L·s)$；k 为速率常数；$[M]$、$[I]$ 分别为单体和引发剂的浓度，mol/L。

---

❶ 第二阶段即低温聚合阶段应严格控制反应温度，一般应保持在45℃以下，待全部产品都不流动时才能升温。

Polymerization rate is directly proportional to the square root of concentration of initiator and first power of the monomer concentration. Normally, k increases with temperature. Therefore, increasing the reaction temperature can increase the rate of polymerization.

Polymerization reaction can be blocked by chain transfer reaction of some compounds and these compounds are known as retarder or inhibitor. The common retarder or inhibitor includes quinone (hydroquinone is easy to act with oxygen or radicals and converts to quinone), nitro compounds and aromatic amines.

Oxygen reacts with reactive free radicals to form non-active free radical. Oxygen is an effective inhibitor for the common free-radical polymerization. In this experiment, methyl methacrylate is regarded as monomer, and test the effect of some factors, such as amount of initiator, oxygen in the air as well as other inhibitor, on the polymerization rate.

3. Instruments and reagents

Instruments: thermostatic water bath, small polymeric tube (with anti-opening rubber stopper), vacuum system, ice-salt bath.

Reagents: methyl methacrylate, azobisisobutyronitrile, hydroquinol.

4. Experimental procedure

(1) Effect of initiator amount on polymerization rate

Take 5 small polymeric tubes with a small rubber stopper ❶ and number the clean tubes in advance. Inject 0 mL, 0.02 mL, 0.10 mL, 0.20 mL and 0.60 mL of methyl methacrylate solution of azodiisobutyronitrilethe (100 mg/mL) into 5 tubes with 1mL syringe, respectively ❷. Distilled methyl methacrylate is added into every polymerization tube, and volume of liquid in each tube is 2 mL. The nozzle is plugged with cleaned rubber stopper ❸ and mix thoroughly (can be performed in liquid mixer). Take a three-way cock connecting a $7^{\#}$ syringe with long needle on its public side, and the other ends are connected to a source of nitrogen and vacuum systems. Needles are inserted into anti-opening rubber stopper in the ice water bath. The solution is degassed by alternately passing nitrogen below the liquid level and evacuating at the level above for 3-5 times. (To prevent the monomer

---

❶ Polymeric tubes can be made in size of ϕ 12 mm × 70 mm with thicker wall. And nozzle should be a taper and tighted with rubber stopper. You can also use a small sealed tube in place. In addition, after evacuating nitrogen in a sealed tube, seal the neck, then polymerize under the predetermined conditions.

❷ Methyl methacrylate solution of azobisisobutyronitrile may be temporarily prepared by laboratory unified, stored in the refrigerator. Initiatior is added to 5 tubes, and the concentration of monomer in each tube is 0%, 0.10%, 0.50%, 1.0% and 3.0%, respectively.

❸ The new rubber stopper is boiling with alkaline water before use, then wash with water to remove dust and other impurities on the surface.

聚合总速率与引发剂浓度的平方根成正比，和单体浓度的一次方成正比。在通常的聚合温度下，常数 k 随温度升高而增大，因而提高反应温度可以增大聚合反应的速率。

某些化合物通过链转移作用可以阻滞或阻止聚合反应的进行，这些化合物被称作缓聚剂或阻聚剂。常用的阻聚剂或缓聚剂有醌类（对苯二酚易与自由基或氧作用而转化为醌类）、硝基化合物和芳香胺类。

氧与活泼的自由基作用后会形成不活泼的自由基，所以氧气对于一般的自由基聚合反应来说，是一种有效的阻聚剂。本实验以甲基丙烯酸甲酯为单体，试验引发剂用量、空气中的氧以及其他阻聚剂对聚合反应速率的影响。

三、仪器与试剂

仪器：恒温水槽；小号聚合管（带反口橡皮管）；抽真空系统；冰盐浴。

试剂：甲基丙烯酸甲酯；偶氮二异丁腈；对苯二酚。

四、实验内容

（1）引发剂用量对聚合反应速率的影响。取 5 支带小号反口橡皮塞的小聚合管❶，预先洗净烘干编上号，用 1mL 注射器注入浓度为 100 mg/mL 的偶氮二异丁腈的甲基丙烯酸甲酯溶液，5 支管中加入量依次为 0 mL、0.02 mL、0.10 mL、0.20 mL、0.60 mL❷。各聚合管中再加入新蒸的甲基丙烯酸甲酯，使各管中液体量均为 2 mL。用干净的小翻口橡皮塞塞紧管口❸，充分混合均匀（可在液体混合器上进行）。取一支三通活塞，在其公用端上接上一支 7 号注射器长针头，另两端分别连接氮气源和抽真空系统。针头从置于冰水浴中的聚合管上的翻口橡皮塞扎入，在液面下通入氮气，在液面上抽真空，交替进行，3~5

---

❶ 小聚合管可做成 ϕ 12 mm × 70 mm 的尺寸，管壁稍厚，管口应有一定锥度并与小号翻口橡皮塞紧配。也可以用小封管代替，在加料并经抽空通氮气处理后，于封管的细颈处封死，在预定的条件下进行聚合。

❷ 偶氮二异丁腈的甲基丙烯酸甲酯溶液可由实验室临时统一配制，配好后置于冰箱中存放。5 个管中加入的引发剂量分别是单体用量的 0%、0.10%、0.50%、1.0% 和 3.0%。

❸ 新购的翻口橡皮塞在使用前要用碱水煮沸并用水洗净，以除去表面粘附的粉尘和其他杂质。

from being drawn out, turn the three-way cock into evacuation system when needle gets out of the liquid, then the three-way nitrogen source take place of evacuation system when the needle is inserted into water). Needle eyes on rubber stopper can be plugged with a little sealant. After completion all above, put the five polymerization tubes into thermostatic water bath at 60 ℃ simultaneously (keep the stopper over the water surface), then observe the changes of liquid viscosity in tubes (speed of rise and disappearance of bubbles generated by shaking becomes significantly slower) and record the time of very sticky (nearly no liquidity) and full no-flow status. Record the results in Table 2.3.

Table 2.3 Record of experimental results

Numbers of samples	Initiator concentration/%	Stiffen time/min	Very sticky time /min
1			
2			
3			
4			
5			

(2) Effect of oxygen in air and inhibitor amount on the polymerization rate.

Take 4 tubes and number them, then add 0.1mL initiator solution to every tube (dosage of initiator is 0.5%) while each is with 1.9 mL methyl methacrylate solution. Add inhibitor hydroquinone to tubes and dosage of 4 tubes is 0%, 0.1%, 0.5%, 1.0%, respectively. Mix the solution evenly in the first 3 polymerization pipes, then treat with nitrogen environment and vacuum treatment followed (1). The last one is plugged in the air, then mix the raw evenly. Put the above four tubes into the water bath at 80 ℃ simultaneously, observe changes of viscosity, record time-t of sticky and very sticky of each polymerization system in the same way.

(3) Effect of temperature on the polymerization rate

Take three same polymerization tubes, and number them. Add 1.9 mL methyl methacrylate and 0.1mL initiator solution into each tube (concentration of monomer is 0.5%). After vacuuming and nitrogen treatment, keep polymering respectively at 50 ℃, 60 ℃, 70 ℃ constantly. Observe and record changes during reaction.

Questions

(1) Describe the effect of initiator concentration and temperature on the polymerization rate qualitatively based on the experimental results.

(2) Please explain the impact of the polymerization inhibitor and the oxygen in the air on the polymerization rate.

次（为防止单体被抽出，应在针头拔出液面后再将三通转向抽空系统，而在三通转向氮气源后再把针头插入到液面下）。橡皮塞上的针眼可用少许密封胶堵上。完成了上述操作后，这 5 支聚合管同时放入已升温至 60℃的恒温水槽中（保持翻口塞在水面之上），观察各管中液体逐渐变黏（经摇动产生的气泡上升与消失的速度显著变慢）、很黏（接近无流动性）和完全不流动的时间。结果记录在表 2.3 中。

表 2.3 实验结果记录表

样品编号	引发剂用量 /%	变黏时间 /min	很黏时间 /min
1			
2			
3			
4			
5			

（2）空气中的氧气及阻聚剂对聚合速率的影响。取 4 支同样的小聚合管另行编号，各加入前述引发剂溶液。引发剂溶液 0.1mL（引发剂用量 0.5% 左右）和甲基丙烯酸甲酯 1.9 mL，然后加阻聚剂对苯二酚，4 支管中的用量分别为 0%、0.1%、0.5%、1.0%。前 3 支聚合管按（1）中要求混合均匀后进行通氮抽真空处理，后一支管就在空气中塞上翻口塞，随即混合均匀。将以上 4 支聚合管同时放入 80℃的恒温水浴，观察其聚合变黏的情况，用类似于表 2.3 的表格记录下各聚合体系变黏和很黏的时间。

（3）温度对聚合反应速率的影响。取 3 支同样的聚合管，编号后各加入甲基丙烯酸甲酯 1.9 mL 和引发剂溶液 0.1 mL（单体量的 0.5%），经仔细抽真空通氮气处理后，分别在 50℃、60℃、70℃的恒温下进行聚合，观察并记录其反应情况。

思考题

（1）根据实验结果定性说明引发剂浓度和反应温度对聚合反应速率的影响。
（2）试说明阻聚剂及空气中的氧对聚合反应速率的影响。

Experiment 3 Suspension polymerization of PMMA for molding powder

1. The purposes

(1) Learn about the mechanism and process characteristics of suspension polymerization.

(2) Be familiar with preparation and function of inorganic suspension stabilizers.

2. Experimental principle

Under the help of dispersant, monomers are dispersed in their own insoluble solvent through vigorously stirring, then suspension polymerization carries out. Kinetics of the reaction is similar to bulk polymerization under similar conditions.

The main feature of the suspension polymerization is that the heat of polymerization can be easily removed and the temperature is easily controlled. Mixture of reaction has good fluidity. Process is easy to handle and molecular weight distribution is uniform. The disadvantage is that it is difficult to produce polymer continuously.

The key to the suspension polymerization is that monomer disperses uniformly and stably. This mainly depends on the ratio of monomer to medium, the dosage and efficiency of dispersant, the shape of mechanical stirrer and stirring speed, etc.

There are two kinds of stabilizerss for suspension polymerization: One is water-soluble polymer, such as gelatin, poly (vinyl alcohol), methyl cellulose and the like; the other is water-insoluble inorganic compound powder, such as calcium carbonate, talc, calcium phosphate and the like. The former is used to increase viscosity of aqueous phase, and the later soaked in water makes the monomer droplets separate from each other. The dispersant should be removed from the final product after polymerization.

PMMA is a copolymer of methyl methacrylate and a minor amount of styrene. Tricalcium phosphate prepared by excessive $CaCl_2$ and Na_3PO_4, under alkaline conditions is chosen as a suspending agent in this experiment. The suspension formed by tricalcium phosphate and water is used directly in the preparation of organic glass molding powder.

3. Instruments and reagents

Instruments: three-neck round bottom flask (250 mL); beaker (100 mL); graduated cylinder (25 mL); mechanical stirrer; condenser tube; watching-glass; glass rod; electric heating jacket (need to control temperature); thermometer; thermometer casing; glass stopper; electric heating jacket; iron support; water bath.

实验三 悬浮聚合法制备有机玻璃模塑粉

一、实验目的
（1）了解悬浮聚合机理及其工艺特点。
（2）熟悉无机悬浮稳定剂的制备及其作用。

二、实验原理
悬浮聚合是在分散剂的作用下，单体依靠剧烈搅拌而分散在其本身不溶的介质中进行的聚合反应。其反应的动力学特征和在相似条件下的本体聚合相仿。

悬浮聚合的主要特点是聚合热容易移去；聚合温度容易控制；反应混合物的流动性好，工艺上易于处理；产物分子量高分布均一。其缺点是难以连续化生产。

实现悬浮聚合的关键是使单体小滴稳定地分散在介质中。这主要取决于单体和介质的用量比、分散剂的用量及效率、机械搅拌器的形状和搅拌速度等条件。

用于悬浮聚合的分散稳定剂有两类：一类是水溶性高分子化合物，如明胶、聚乙烯醇、甲基纤维素等；另一类是难溶于水的无机化合物粉末，如碳酸钙、滑石粉、磷酸钙等。前者主要用以增加水相的黏度，后者在被水浸润后使单体的液滴互相分离。分散剂在聚合后都应当从体系中除去。

有机玻璃模塑粉是甲基丙烯酸甲酯和少量苯乙烯的共聚物。本实验用磷酸三钙 $[Ca_3(PO_4)]_2$ 作为悬浮剂（由过量的 $CaCl_2$ 与 Na_3PO_4 在碱性条件下制得），形成的磷酸三钙悬浮液可以直接用于有机玻璃模塑粉的制备。

三、仪器与试剂
仪器：250 mL 三口反应瓶；100 mL 烧杯；25 mL 量筒；搅拌器；冷凝管；表面皿；玻璃棒；电加热套（需控制温度）；温度计；温度计套管；玻璃塞；电加热套；铁架台；水浴锅。

Reagents: Methyl methacrylate; styrene; benzoyl peroxide; stearic acid; $CaCl_2$; Na_3PO_4; NaOH; concentrated hydrochloric acid; 1% $AgNO_3$ solution.

4. Experimental procedure

(1) Preparation of suspension agent

30 mL of 1.0% $CaCl_2$ solution (about 0.001mol) is added to the 250 mL three-neck round bottom flask equipped with condenser tube and mechanical stirrer. Keep heating until the solution temperature reaches to 95°C. At the same time, put 30 mL mixture aqueous solution of 1.0% Na_3PO_4 (about 0.002 mol) and 0.25% NaOH (about 0.002 mol) to a 100 mL beaker ❶. The latter is heated on an electric heating jacket until it is nearly boiling, then dropped into the three-necked flask with $CaCl_2$ solution under stirring within five minutes. Then incubate the suspension at 95°C for 0.5 h. Cool it and standby ❷.

(2) Polymerization

Add 80 mLof distilled water to the cool suspension agent and mix them thoroughly. The diluted suspension is bubbled with nitrogen to remove the oxygen. The polymerization should go on under a nitrogen atmosphere.

25 mL of methyl methacrylate (0.25 mol), 5.5 mL of styrene, 0.05 g of stearic acid and 0.2 g of benzoyl peroxide are added to a 250 mL beaker. The oil phase is dissolved by agitation and poured into the three-neck bottle. Make the system dispersed evenly by agitation. Meanwhile, react for 1h at 90°C. Then heat from 110°C to 115°C, reaction is going on for 1h ❸.

(3) Post treatment

The final product can be obtained by acid washing, water washing, filtration and drying. Briefly, decant the aqueous layer when the reaction mixture is cooled. 2 mL of concentrated hydrochloric acid is added to make solution be neutral and remove alkaline calcium. Then wash the crude product with distilled water for 4~5 times, until no Cl^- ($AgNO_3$ solution is available for detection) can be detected in the filtrate. After filtering and washing, the precipitate was spread out and placed in a large culture dish, and dried in vacuum at 80 °C. Weigh and calculate the yield.

---

❶ For convenience, $CaCl_2$, NaOH, Na_3PO_4 and other solutions can be prepared by the laboratory in whole. Process to make solution: 10 g $CaCl_2$ is dissolved in 1L distilled-water; 10 g Na_3PO_4 and 2.5 g NaOH are dissolved in 1L-distilled water.

❷ The suspension is a stable milky-white suspension. It can not be used if there is precipitation occurs. A small amount of suspension was diluted with distilled water for 10 times without precipitation, indicating that it can be used. The suspension agent should be used promptly and cannot stay too long.

❸ The product is a copolymer with a few styrene segments in the main chain of polymethylmethacrylate. It is generally considered that the processed product has ideal physical and mechanical properties when the molecular weight of the copolymer reachs 130~150 kDa.

试剂：甲基丙烯酸甲酯；苯乙烯；过氧化苯甲酰；硬脂酸；$CaCl_2$；Na_3PO_4；NaOH；浓盐酸；1% $AgNO_3$ 溶液。

四、实验内容

（1）悬浮剂的制备

在装有冷凝管和搅拌器的 250 mL 三口反应瓶中，加入 1.0% 的 $CaCl_2$ 溶液 30 mL（约 0.001mol），加热至 95℃。同时在 100 mL 小烧杯中加入含有 1.0% 的 Na_3PO_4（约 0.002 mol）和 0.25%NaOH（约 0.002 mol）的水溶液 30 mL❶。后者在电加热套上加热至近沸后，于搅拌条件下滴加到盛有 $CaCl_2$ 溶液的三口瓶中，在 5 min 内加完。然后 95℃下保温 0.5h，冷却备用❷。

（2）聚合

在经冷却至室温的上述悬浮剂乳液中加入 80 mL 蒸馏水，通入氮气以除去瓶内氧气。整个反应均需在氮气气氛下进行。

在 250mL 的烧杯中，加入甲基丙烯酸甲酯 10 mL（0.1mol）、苯乙烯 2.2mL、硬脂酸 0.02 g 和过氧化苯甲酰 0.08 g。搅拌使之溶解混合均匀后加到三口反应瓶中。开动搅拌使体系分散均匀。同时快速升温至 80℃，反应 1h 后再升温至 90℃反应 1h，然后升温至 110~115℃，继续反应 1h❸。

（3）后处理

反应混合物需经酸洗、水洗、过滤和干燥等处理后方能得到最终的产品。待反应混合物冷却后，倾去上面的水层，加入 2 mL 浓盐酸以中和其碱性并除去钙盐。再用水洗 4~5 次，至无 Cl^- 存在（可用 $AgNO_3$ 溶液检查）。沉淀物经抽滤洗涤后，置于大号培养皿中，摊开，在 80℃下真空干燥，称重，计算产率。

---

❶ 为方便起见，$CaCl_2$、NaOH、Na_3PO_4 等溶液可由实验室统一配好。配制方法：10 g $CaCl_2$ 溶于 1L 蒸馏水中，10 g Na_3PO_4 和 2.5 g NaOH 溶于 1L 蒸馏水中。

❷ 反应得到的悬浮剂应为稳定的乳白色悬浊液。若出现沉淀则不能使用。取少量悬浊液用蒸馏水稀释 10 倍，仍不发生沉淀表明可以使用。悬浊液配好后要及时使用，不宜久置。

❸ 产物是在聚甲基丙烯酸甲酯主链中掺有少数苯乙烯链节的共聚物，一般认为共聚物的分子量达到 13 万~15 万时，加工后的产品才具有较理想的物理和机械性能。

Questions

(1) Taking polyvinyl alcohol and tricalcium phosphate as examples, try to discuss the similarities and differences between the principle of suspension effect of hydrophilic organic suspending agent and water-insoluble inorganic suspending agent.

(2) What is the effect of stearic acid on the final properties of the product?

(3) Is it necessary to remove the inhibitor in methyl methacrylate? Why?

Experiment 4   Depolymerization of Poly(methyl methacrylate)

1. The purposes

(1) Know about the cleavage reaction of polymers by pyrolysis of plexiglass.

(2) Learn how to purify monomer by steam distillation.

2. Experimental principle

Degradation reaction is a reaction that the polymer chain is split into shorter ones. Take degradation responses in natural polymer for example, amino acids are from the protein degradation and glucose are from starch or cellulose. Synthesize new polymer and the synthetic polymers can be applied to recover some monomers following the rules, such as the synthesis of cyanoacrylate adhesive (e.g. 502 glue) composed of α-cyanoacrylate active monomer, etc.

Thermal stability of the polymer, the pyrolysis rate and yield of monomer are all closely related to chemical structure of the polymer. Degradation of the main chain has three cases: The main chain breaks at any point, which is random degradation; monomers come off from the end of main chain constantly, which is depolymerization reaction; the synergies of the two reactions. Experimental facts show that: Degradation of polyethylene is random; reactions of poly (methyl methacrylate), poly (α-methyl styrene), poly (methacrylic acid) and polytetrafluoroethylene are depolymerization; case of polystyrene is between the two above.

The main depolymerization product of poly (methyl methacrylate) (PMMA) is its monomer methyl methacrylate. In addition, there are a small amount of oligomers, methacrylic acid and some thermal decomposition products coming from plasticizer of phthalates in the product. In order to remove these impurities before distillation (otherwise, the temperature in the distillation bottle will be too high, which will cause the polymerization in the distillation process), it is necessary to extract pre-products generated by pyrolysis of organic glass by water vapor distillation.

Water vapor distillation is one of the most commonly used methods for the separation of water

思考题

（1）以聚乙烯醇和磷酸三钙为例，讨论亲水有机悬浮剂和水不溶的无机悬浮剂两者的悬浮作用原理有何异同。

（2）加入硬脂酸将对产物最终性能起什么作用？

（3）甲基丙烯酸甲酯是否需要去除阻聚剂？为什么？

实验四　有机玻璃的解聚

一、实验目的

（1）通过有机玻璃的热裂解，了解高聚物的裂解反应。

（2）学习用水蒸气蒸馏法纯化单体。

二、实验原理

降解反应是指高分子链被分裂成为较短链的反应过程。降解反应用于天然高分子可由蛋白质制取氨基酸，从淀粉或纤维素制取葡萄糖；应用于合成高分子可以回收某些单体，制取新型聚合物，如瞬干胶（如502胶）α-氰基丙烯酸酯类等活性单体的制备。

聚合物的热稳定性、裂解速度以及单体的收率是和聚合物的化学结构密切相关的。聚合物主链降解有3种情况：主链上任意点发生断裂的无规降解；单体有规则地从主链末端不断脱落下来的解聚反应，以及上述两种反应的协同作用。实验事实表明：聚乙烯发生无规降解，聚甲基丙烯酸甲酯、聚α-甲基苯乙烯、聚甲基丙烯酸及聚四氟乙烯则发生解聚反应；聚苯乙烯的情况介于这二者之间。

有机玻璃的解聚产物主要是其单体——甲基丙烯酸甲酯，此外还有少量低聚物、甲基丙烯酸以及作为增塑剂添加进去的邻苯二甲酸酯类的热分解产物。为在精馏前除去这些杂质（否则会引起精馏瓶中温度过高，造成精馏过程中的再聚合），需要对有机玻璃裂解的初产物进行水蒸气蒸馏。

水蒸气蒸馏是分离和纯化有机化合物常用的方法之一，要求被提纯的物质

insoluble or slightly soluble compounds. or water insoluble compounds, which requires the purified substance must have a certain vapor pressure at 100 ℃. Methyl methacrylate is nearly insoluble in water and vapor pressure is close to 100 kPa at 100 ℃. Therefore, methyl methacrylate can be separated by vapor distillation.

3. Instruments and reagents

Instruments: round bottome flask (250 mL); three-neck round bottom flask (250 mL); steam generator; separating funnel; condenser; electric heating-jacket.

Reagents: Plexiglass trim; concentrated H_2SO_4 solution; saturated Na_2CO_3 solution; saturated NaCl solution; anhydrous Na_2SO_4.

4. Experimental procedure

(1) Cracking apparatus is shown in Figure 2.1. Weigh 50 g of plexiglass scraps and put them into a 250 mL three-neck round bottom flask ❶. Heat flask to 200 ℃ with heating units and control the speed of heating to ensure lysate dropwise outflows. The de-polymerization reaction is nearly completed when the inner temperature reaches to 350 ℃, then stop heating and remove the receiving flask. Finally weigh distillate and calculate the yield.

(2) The above obtained lysate is extracted by steam distillation, and its cracking apparatus is shown in Figure 2.2 ❷. Bottle A as shown is the steam generator tube while long glass tube inserted is for safe. Bottle B is tilted appropriately to reduce the possibility of liquid pouring out of bottle.

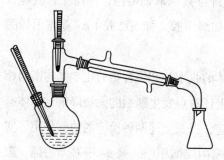

Figure 2.1 Cracking apparatus of plexiglass

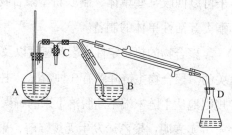

Figure 2.2 Steam distillation apparatus of cleavage product

A—steam generating bottle; B—retort;
C—free-clip; D—receiving flask

---

❶ To facilitate heat-transfer, plexiglass scrap is pulverized using hammer or file.

❷ In the steam distillation of a small amount of pyrolysis products, the steam generator can not be used. Only a certain amount of distilled water is added into the product and then heated and evaporated out. If necessary, it can be steamed out by adding water several times.

在100℃时必须具有一定的蒸气压，在水中不溶或微溶。甲基丙烯酸甲酯基本上不溶于水，在100℃时其蒸气压已接近100 kPa，故可用水气蒸馏的方法分离出甲基丙烯酸甲酯单体。

三、仪器和试剂

仪器：250 mL 圆底烧瓶；三口瓶；水蒸气发生装置；分液漏斗；冷凝管；电热套。

试剂：有机玻璃边角料；浓 H_2SO_4；饱和 Na_2CO_3 溶液；饱和 NaCl 溶液；无水 Na_2SO_4。

四、实验内容

（1）裂解装置如图 2.1 所示。称取 50 g 有机玻璃边角料放入 250 mL 三口瓶中❶，用电热套加热至200℃，控制升温速度，以保证裂解产物逐滴流出。当瓶内温度达350℃时，解聚反应已接近完全，此时应停止加热，取下接收瓶，称量馏出物，计算其收率。

（2）将上述的裂解粗产物进行水蒸气蒸馏，其装置如图 2.2❷所示。图中瓶A为蒸气发生器，其中插入的一长玻璃管为安全管。B瓶适当倾斜放置，以减小瓶内液体被带出的可能性。

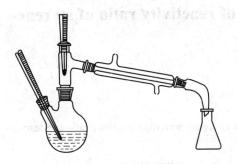

图 2.1 有机玻璃解聚装置

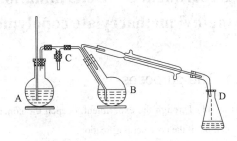

图 2.2 裂解产物的水蒸气蒸馏装置
A—水蒸气发生瓶；B—蒸馏瓶；C—自由夹；
D—接收瓶

---

❶ 为便于传热，有机玻璃边角料要进行粉碎处理，可用锤子敲碎或用锉刀锉成碎屑。

❷ 在对少量裂解产物进行水蒸气蒸馏时，也可以不用水蒸气发生器，只需要在产物中加入一定量的蒸馏水后加热蒸出，需要时也可以分多次加水蒸出。

Put the primary distillate into bottle B, and heat bottle A resulting in water vapor flowing from bottle A to bottle B. Heat bottle B with electric heating-jacket at the same time ❶. The mixed vapor is completely condensed in the condenser tube, then end off distillation apparatus when the collection of the distillate do not contain the monomer droplets any longer. Open the free-clip in the lower end of tees used to connect A, B bottles (to prevent residue from bottle B back to the bottle A), then stop heating bottle A.

(3) The steam distillate is placed in a 150 mL of separating funnel, then discards water layer and add 1~2 mL of concentrated H_2SO_4 (3%-5% by mass of the monomers) to wash away the unsaturated hydrocarbons and alcohols impurities in monomer. Wash twice to remove most of the acid with 20 mL of distilled-water, then wash once with saturated Na_2CO_3 solution and finally with saturated brine washing solution to be neutral. Dry with anhydrous Na_2SO_4 for further purification.

(4) The dried monomer is extracted by vacuum distillation ❷, and collect the fraction under constant pressure. The samples are analyzed by infrared spectroscopy or refractive index, and the results were compared with standard spectra or standard values.

Questions

(1) What are the effects of pyrolysis temperature and pyrolysis speed on the product quality?

(2) Please write the reaction formula of concentrated H_2SO_4 with impurities of alkenes and alcohols in monomer, and explain how they are removed?

Experiment 5 Determination of reactivity ratio of styrene-methyl methacrylate copolymer

1. The purposes

(1) Through the experiment, deepen the concept of copolymerization reactivity ratio and learn a method of the polymer purification.

(2) Analyze the composition of styrene-methyl methacrylate copolymer by UV spectrophotometry.

---

❶ Control the heating time and heating rate of bottle B carefully to make velocity of steam entering is approximately equal to that of the steam. Don't let liquid get out of bottle B.

❷ The degree of vacuum is not too high, generally within 15-30 kPa, to ensure that the water vapor is completely condensed when the methyl methacrylate is extracted by vacuum distillation. The boiling range of methyl methacrylate is 50 ~ 65 °C at this pressure.

将初馏物（约35 mL）加入B瓶中，加热A瓶，使产生的水蒸气进入B瓶，并同时用电热套加热B瓶❶。使混合蒸气在冷凝管中全部冷凝下来，当收集的馏出液不再含有单体珠滴时，结束蒸馏操作。结束时，先打开连结A、B瓶间的三通管下端的自由夹（防止B瓶中的残留物倒吸到A瓶中去），再停止对A瓶的加热。

（3）将水蒸气馏出物置于150 mL分液漏斗中，分去水层，加入浓H_2SO_4 1~2mL（单体质量的3%~5%），以洗去单体中不饱和烃及醇类杂质。再用20 mL蒸馏水洗2次以除去大部分酸，然后用饱和Na_2CO_3溶液洗一次，最后用饱和食盐水将单体洗至中性。用无水Na_2SO_4干燥以备进一步精制。

（4）将上述干燥后的单体进行减压蒸馏❷，收集一定压力下的某一馏分。样品作红外分析或折射率测定，将结果与标准谱图或标准值进行比较。

思考题

（1）裂解温度的高低及裂解速度的大小对产品的质量有什么样的影响？

（2）写出浓H_2SO_4与单体中的杂质烯和醇的反应式，说明它们是如何被除去的？

实验五　苯乙烯-甲基丙烯酸甲酯共聚反应竞聚率的测定

一、实验目的

（1）通过实验，加深对共聚反应中竞聚率概念的理解，学习一种纯化聚合物的方法。

（2）学习用紫外分光光度法测定苯乙烯-甲基丙烯酸甲酯共聚物的组成。

---

❶ 仔细控制B瓶的加热时间和升温速度，使蒸气进入的速度与蒸出的速度大致相等。注意不可将B瓶中的液体蒸干。

❷ 在减压蒸馏甲基丙烯酸甲酯时，为使其蒸气被充分冷凝下来，真空度不宜太高，一般用水泵控制压力在15~30 kPa，这时甲基丙烯酸甲酯的沸点范围是50~65℃。

2. Experimental principle

Copolymerization reaction system refers to the reaction of two or more than two monomers getting together. The product of the reaction is called copolymer. In this experiment, the copolymerization of styrene and methyl methacrylate was carried out with azodiisobutyronitrile as initiator.

The reactivity ratio of monomer is the ratio of the reaction rate constant between the end of free radical chain (active center) of this monomer and the reaction rate constant of the same kind of monomer to that of another monomer. The reactivity ratios represents that the active center has different tendencies to choose two monomer. When the two monomers are copolymerized, their reactivity ratios can be expressed as r_1 and r_2 respectively.

According to the knowledge of polymer chemistry, conversion rate of the polymerization reaction is less than 10% and if the structural units in the polymer chain have no effect on the polymerization reaction, there establish the following formula:

$$r_2 = r_1 \frac{[M_1]^2}{[M_2]^2} \frac{[m_2]}{[m_1]} + \frac{[M_1]}{[M_{21}]} \left(\frac{[m_2]}{[m_1]} - 1 \right) \tag{1}$$

Where $\frac{[M_1]}{[M_2]}$ is The mol ratio of the first monomer to second monomers in the feed. Where $\frac{[m_2]}{[m_1]}$ is The mol ratio of structure unit of the second monomers in the copolymer to the first monomer

According to Lambert-Beer's law:

$$E = \lg \frac{I_0}{I} = kCL \tag{2}$$

Where E is extinction or absorption. Where I_0 is incident intensity. Where I is transmitted light intensity. Where k is extinction coefficient. Where C is concentration of light-absorbing substance. Where L is absorption cell thickness.

Make use of characteristic absorption of benzene at 261 nm in the ultraviolet region to measure the absorbance of pure polystyrene solution and pure poly (methyl methacrylate) solution (in a chloroform solution as a reference) at a specific wavelength (at 261nm, where styrene has larger absorption and methyl methacrylate (MMA) has no obvious absorption) and convert into normalized vaule of uptake[❶]. You can make the whole work curve of styrene and its weight percentage is 0%-100%. Determine the normalized absorption of polymer obtained under different circumstances from the working curve. Then find the weight percentage of styrene unit and calculate the value of $[m_2]/[m_1]$, then get r_1, r_2 and error by

---

❶ Normalized absorption is equivalent to the absorption at a certain concentration. If the specified reference concentration is C, the actual concentration is $C \pm \Delta C$, the measured extinction value is E' and the normalized absorbance value is c'. All measured absorption values can be compared and calculated only after normalization.

二、实验原理

共聚反应系指两种或两种以上的单体在一起进行聚合的反应,反应的产物称为共聚物。本实验用苯乙烯和甲基丙烯酸甲酯两种单体,以偶氮二异丁腈(AIBN)为引发剂进行共聚合反应。

单体的竞聚率是指这种单体的自由基链末端(活性中心)和同种单体反应的反应速率常数与其和另一种单体反应的速率常数的比值,它表征着该活性中心对两种单体的不同选择聚合趋势。当两种单体进行共聚时,它们的竞聚率可分别表示为 r_1 和 r_2。

根据高分子化学有关知识,在聚合反应的转化率低于 10% 并且假定除了链末端的自由基单元,聚合物链中的其他结构单元对聚合反应无影响时,有下式成立:

$$r_2 = r_1 \frac{[M_1]^2}{[M_2]^2} \frac{[m_2]}{[m_1]} + \frac{[M_1]}{[M_{21}]} \left(\frac{[m_2]}{[m_1]} - 1 \right) \quad (1)$$

式中,$\frac{[M_1]}{[M_2]}$ 为投料中第一种单体对第二种单体的摩尔比;$\frac{[m_2]}{[m_1]}$ 为共聚物中第二种单体的结构单元对第一种单体结构单元的摩尔比。

根据朗伯-比尔(Lambert-Beer)定律有:

$$E = \lg \frac{I_0}{I} = kCL \quad (2)$$

式中,E 为消光或吸收;I_0 为入射光强;I 为透射光强;k 为消光系数;C 为吸光物质的浓度;L 为吸收池厚度。

利用苯环在紫外区约 261 nm 处的特征吸收,测出纯聚苯乙烯溶液和纯聚甲基丙烯酸甲酯溶液在特定波长(本实验为 261nm,这时苯乙烯有较大吸收而甲基丙烯酸甲酯无明显吸收)下的吸收值(以溶剂氯仿为参比),换算成归一化吸收[1]后,即可作苯乙烯质量分数 0%~100% 的全部工作曲线。测定在不同配料比下所获得的共聚物的归一化吸收,从工作曲线上可以查出它们相应的含苯乙烯结构单元的质量分数,进而可以算出值 $[m_2]/[m_1]$,再根据每一组 $[M_1]/$

---

[1] 归一化吸收是指相当于某一基准浓度下的吸收。若指定基准浓度为 C,实际溶液浓度为 $C \pm \Delta C$,实测消光值为 E',则归一化吸收值为 C'。各个测定的吸收值只有归一化后方可进行比较和运算。

the method of graphing or statistical according to each group value of $[M_1]/[M_2]$ and the value of $[m_2]/[m_1]$.

3. Instruments and reagents

Instruments: Super constant temperature water bath; vortex mixer; polymerization stopper tube (50mL); syringe needle; volumetric flask; pipette; beaker; 751G spectrophotometer; nitrogen cylinder.

Reagents: Freshly distilled styrene and methyl methacrylate; recrystallized azobisisobutyronitrile; chloroform; petroleum ether (boiling range: 60-90 ℃).

4. Experimental procedure

(1) Label the seven cleaned and dried polymerization tubes (50 mL) and add 30 mg of azobisisobutyronitrile to every tubes.

(2) According to the dosage in Table 2.4, accurately measure the monomer with a pipette and add it to the corresponding polymerization tube. Plug the mouth of tube ❶ and mix thoroughly in a vertex mixer, then the initiator is dissolved in solution and the two monomers are mixed uniformly.

Table 2.4 Experimental recipe

Number	1	2	3	4	5	6	7
Styrene/mL	30	25	20	15	10	5	0
Methyl methacrylate/mL	0	5	10	15	20	25	30

(3) Put the above polymeric tubes into an ice-water bath. Then keep cooling for 10 min. Insert a long $7^{\#}$ syringe needle into the tube from the rubber stopper. Evacuate with nitrogen alternately by three-way cock which connected to $7^{\#}$ syringe needle and repeat three times (when evacuating, the needle shall be above the liquid level; when N_2 bubbling, the needle shall be below the liquid level. Be careful not to draw monomer into the needle tube). Pull the needle from the tube in the nitrogen circumstance. Then remove the polymerization tube from ice-water bath circumstance. Use about 15 cm-length wire to fasten bottle mouth and mix again in the mixer.

(4) When temperature of polymerization tube rises to room temperature, use thin wire to hang polymerization tube over the super constant temperature water bath at 70 ℃ (Stopper is higher than the surface about 1 cm for 30-60 min. When the reactant becomes sticky (bubbles come from the bottom when shaking polymerization tube, and rate of rise and disappearance are relatively slow), remove the

---

❶ Stoppers are washed with water before use, then boil it with 5% NaOH solution. Finally wash with distilled-water for standby. Pay attention to avoid contact between monomer and rubber stopper after adding monomer.

$[M_2]$ 和 $[m_2]/[m_1]$ 的值，用作图法或者用统计方法得出 r_1，r_2 及其误差。

三、仪器和试剂

仪器：超级恒温水浴槽；漩涡混合器；配有大号翻口塞的聚合管（50 mL）；注射器针头；容量瓶；吸量管；大烧杯；751G 型分光光度计；氮气瓶。

试剂：新蒸苯乙烯和甲基丙烯酸甲酯；重结晶过的偶氮二异丁腈；氯仿，沸程为 60~90℃的石油醚。

四、实验内容

（1）将 7 支容量为 40 mL，带有翻口橡皮塞的聚合管洗净、烘干、标上号码，各称入 30 mg 偶氮二异丁腈。

（2）按表 2.4 中的用量，用吸量管准确地量取单体，加入相应的聚合管中。塞上翻口塞❶，在液体混合器上充分混合，使引发剂溶解并使两种单体均匀混合。

表 2.4　实验配方表

聚合管编号	1	2	3	4	5	6	7
苯乙烯用量 /mL	30	25	20	15	10	5	0
甲基丙烯酸甲酯用量 /mL	0	5	10	15	20	25	30

（3）准备好一只冰水浴烧杯，将上述聚合管放入其中，冷却 10 min 后，从翻口塞上插入一支长的 7# 注射器针头，通过连接于长针头的三通活塞，轮流进行抽空充 N_2，重复 3 次（针头在液面上抽空，在液面下通 N_2，注意不要将单体抽入针管内）。充 N_2 下拔出针头，同时从冰水浴中取出聚合管，用一根 15cm 长的细铁丝扎紧翻口塞，再次在混合器上混匀。

（4）待聚合管恢复到室温后，利用扎管口的细铁丝将聚合管吊挂在 70℃ 的超级恒温水浴中（翻口塞要高出水面 1 cm 左右），反应 30~60 min，当反应物具有一定黏度（摇动聚合管时从底部产生气泡，且上升和消失都比较缓慢）时，取出聚合管（7 支聚合管不必同时取出，可以视反应情况先后取出）。将

---

❶ 翻口橡皮塞在使用前应用清水洗净，再用 5%NaOH 溶液煮沸，最后用蒸馏水洗净备用。注意装入单体后应尽量避免单体和橡皮塞接触。

polymerization tube (7 polymerization tubes do not need to be removed at the same time, depending on the circumstances). The samples are poured into a large beaker with 200 mL of petroleum ether whose boiling point ranges from 60 to 90 ℃ (methanol can also be used) to get the precipitated polymer. In order to avoid the precipitate from sticking to the glass rod, the reaction products should be evenly dispersed in the precipitant without stirring (available in dropper instillation), standing for a few minutes. Then cut the pellet into pieces with small scissors or with a glass rod, and soak for several hours in the precipitant. Pour out of the top precipitant and transfer the polymer to a 50 mL beaker. After most precipitant is volatilized, weigh up the wet weight of resultant (generally should be 1-2 g). Finally, 1g sample is left in the beaker.

(5) 5-10 mL of chloroform is added to 7 small beakers which containing 1g of polymer while keep stirring until completely dissolved, then pour the polymer solution into the flask slowly with 150 mL petroleum ether under stirring. Wait the precipitation falling to the bottom of the beaker, filtrate or pour out supernatant immediately [1]. Squeeze the polymer with a glass rod, wash twice with a small amount of petroleum ether, and then transfer precipitation into a clean weighing bottle. When most of the petroleum ether evaporated, place the samples into vacuum oven, drying 20h at 60 ℃.

(6) The concentration of copolymer suitable for UV analysis is generally 0.5 mg/mL. Take 15.0 mg of dried samples into a 25 mL volumetric flask (the samples are placed in the bottom of the bottle) and dissolve it using 15 mLof chloroform, then dilute to the scale with chloroform before use and shake for UV analysis.

(7) Determine the values of pure polystyrene solution and pure poly (methyl methacrylate) solution absorption in the UV region between 240-290nm with 751G UV-visible spectrophotometer [2]. Measure one point every 5 nm and get one point every 1nm when it come to the maximum absorption value. Each point is repeated measuring for three times, then take the average values and find out the point that is the maximum absorption peak of styrene. The working curve is drawn according to the absorption value of methyl methacrylate and styrene solution at that point.

(8) Determine the absorption values of each copolymer sample at the point where the styrene has the maximum absorption (261nm), then convert into normalized value and find out the concentration of styrene in each sample (mg/mL) from the working curve. Use data processing and graphing or statistical

---

[1] The density of chloroform is high, if you do not separate the precipitate out in time, precipitation will be dissolved again due to the large concentration of chloroform in the bottom of the flask.

[2] If you are using UV-240 spectrophotometer that can scan automatically and data processing, it can greatly simplify the experimental procedure. The instrument can give the value of weight percent of each styrene structure units directly.

各样品分别倒入盛有 200 mL 沸程为 60~90℃的石油醚（也可用甲醇）的大烧杯中，聚合物随即沉淀出来。为避免沉淀物粘结成团而缠绕在玻璃棒上，应在不加搅拌的情况下先将反应产物均匀分散于沉淀剂中（可用滴管滴入），静置数分钟后，用小剪刀将片状沉淀物剪碎，或用玻璃棒将其捣碎，在沉淀剂中搅拌后浸泡数小时，倾出上层沉淀剂。而后将聚合物按管号分别转移到 50 mL 小烧杯中。待样品中的沉淀剂大部分挥发后，称重所得各个样品湿重（一般应为 1~2 g）。留下 1g 样品存在小烧杯中。

（5）在 7 只各盛有 1g 聚合物的小烧杯中分别加入分析纯氯仿 5~10 mL，搅拌使之完全溶解，然后在搅拌下将聚合物溶液慢慢倒入盛有 150 mL 石油醚的烧杯中，待沉淀积于杯底，立即进行抽滤或倾出上层清液❶。用搅拌棒压干聚合物，再用少量石油醚洗涤沉淀 2 次，随后将沉淀移入干净的称量瓶中，待大部分石油醚挥发之后，将样品放入真空烘箱中，于 60℃进行真空干燥 20 h 以上。

（6）适于进行紫外分析用的共聚物浓度一般为 0.5 mg/mL。称取 15.0 mg 干燥样品，放入 25 mL 容量瓶中（应将样品全部放在瓶底），先用 15 mL 分析纯氯仿溶解，临用前用氯仿稀释至刻度，摇匀后供紫外分析用。

（7）以氯仿为参比，在 751G 型可见紫外分光光度计上❷测定纯聚苯乙烯溶液和纯聚甲基丙烯酸甲酯溶液在紫外区 240~290 nm 间的吸收值，每隔 5 nm 测一点，在最大吸收值附近每隔 1 nm 测一点，每点重复测三次取平均值，找出聚苯乙烯最佳吸收峰位置。以两者在该点的吸收值为依据画出工作曲线。

（8）用同样的方法测定各共聚物样品在聚苯乙烯最大吸收峰（261nm）处的吸收值，换算成归一化吸收后，从工作曲线上查出各样品中所含苯乙烯结均

---

❶ 氯仿比重较大，若不及时将沉淀物分离出去，氯仿会在杯底形成较高浓度而使沉淀再度溶解。

❷ 若采用像 UV-240 的能自动扫描和进行数据处理的紫外分光光度计，则可以大大简化这一实验步骤，仪器能直接给出每一样品中苯乙烯结构单元的质量分数数据。

method (least squares method) ❶ to calculate r_2, r_1 value.

Questions

(1) According to the reasonable assumption, write out derivation (1), and explain why the experimental results are more accurate from the theoretical considerations in the case of low conversion.

(2) What's the effect of temperature on the reactivity of the competition?

(3) Analyze the composition of the copolymer of styrene and methyl methacrylate, and know what methods do you have to produce it? Please explain briefly.

---

❶ Typical mapping method:

If
$$a = \frac{[M_1]^2}{[M_2]^2} \times \frac{[m_2]}{[m_1]} \tag{3}$$

$$b = \frac{[M_1]}{[M_{21}]} \left(\frac{[m_2]}{[m_1]} - 1 \right) \tag{4}$$

Then
$$r_2 = r_1 a + b \tag{5}$$

Calculate two values corresponding to r_2, r_1 arbitrary values based on the assumption, then r_2-r_1 in diagram will form a straight line with the equation (5). So, a, b values have a corresponding line for each group. These lines in the diagram will form a cross (the lines should be referred to a point in theory), do the inscribed circle, values of r_2 and r_1 are the coordinates of the center, and size of the inscribed circle radius can be approximately regarded as error of size.

According to Statistical methods, a group values of a, b in line with the linear equation (5) Can be expressed as follows:

$$r_1 = (\Sigma a \Sigma b - n \Sigma ab)/D \tag{6}$$

$$r_2 = \bar{b} + r_1 \bar{a} \tag{7}$$

Where, n- The numbers of a or b, $D = n\Sigma a^2 - (\Sigma a)^2$, $\bar{a}$、$\bar{b}$—Average of a, b

And the statistical errors of r_1, r_2---$S(r_1)$, $S(r_2)$

$$S(r_1) = [nS_0/(n-2)D]^{1/2} \tag{8}$$

$$S(r_2) = [S_0 \Sigma a^2/(n-2)D]^{1/2} \tag{9}$$

Here, $S_0 = \Sigma b^2 - [(\Sigma b)^2/D] + r_1[\Sigma ab - \Sigma a \Sigma b/n]$

In fact, it is convenient to program properly, or to calculate and act directly on the microprocessor.

单元的浓度（mg/mL），进行数据处理后，用作图法或统计法（最小二乘法）[1] 求出 r_2、r_1 值。

思考题

（1）试根据合理假设，推导出式（1），并说明为什么转化率越低，则从理论上考虑认为实验结果将越精确。

（2）温度对竞聚率有怎样的影响？

（3）对苯乙烯和甲基丙烯酸甲酯共聚物的组成分析，你还能提出哪几种方法？简单说明之。

---

[1] 典型的作图方法：

若令
$$a = \frac{[M_1]^2}{[M_2]^2} \times \frac{[m_2]}{[m_1]} \tag{3}$$

$$b = \frac{[M_1]}{[M_{21}]} \left(\frac{[m_2]}{[m_1]} - 1 \right) \tag{4}$$

则前面的方程（1）就转换成直线方程：
$$r_2 = r_1 a + b \tag{5}$$

对于每一组的 a、b 值（相应于不同的配料比），假设 2 个任意的 r_1 值，由此计算 2 个相应的 r_2 值，在 r_2 对 r_1 的直角坐标图中可得到符合方程（5）的一条直线。这样，对于每一组的 a、b 值就有一条相应的直线，这些直线在图中会形成一个交区（理论上应交于一点），作这一交区的内接圆，圆心所在点的坐标即是所求 r_2、r_1 值，而内接圆半径的大小可近似看作误差的大小。

对于一组符合直线方程（5）的 a、b 值，应有：
$$r_1 = (\Sigma a \Sigma b - n \Sigma ab) / D \tag{6}$$
$$r_2 = \bar{b} + r_1 \bar{a} \tag{7}$$

式中，n 为 a 或 b 值的个数；$D = n\Sigma a^2 - (\Sigma a)^2$；$\bar{a}$、$\bar{b}$ 分别为 n 个 a、b 的平均值。r_1、r_2 的统计误差 $S(r_1)$、$S(r_2)$ 分别为：
$$S(r_1) = [nS_0 / (n-2)D]^{1/2} \tag{8}$$
$$S(r_2) = [S_0 \Sigma a^2 / (n-2)D]^{1/2} \tag{9}$$

式中，$S_0 = \Sigma b^2 - [(\Sigma b)^2 / D] + r_1 [\Sigma ab - \Sigma a \Sigma b / n]$

实际上，编制适当的程序，在微处理机上直接进行上述运算和作图是极为快速方便的。

Experiment 6　Synthesis of unsaturated polyester resin and glass fiber reinforced plastics

1. The purposes

(1) Know the principles and methods of controlling polymerization degree of linear polyester.

(2) Master the experimental skills to prepare unsaturated polyester and glass fiber reinforced plastics.

2. Experimental principle

　　Step polymerization is one of the important methods to synthesize polymer materials. Conventional polymer materials such as polyester, nylon, polyurethane and phenolic resin, and high-performance polymer materials such as polycarbonate, polysulfone, polyphenylene ether and polyimide are all synthesized by step polymerization. The step polymerization is carried out through the chemical reaction of functional groups between different substances, which can be monomers or molecular chains with different degrees of polymerization. The molecular weight of step polymerization increases with the degree of functional group reaction. Only at a very high degree of reaction can obtain high molecular weight polymer. According to the reaction types, it can be divided into condensation polymerization, step-by-step addition polymerization and oxidative coupling polymerization, etc. Step polymerization can be performed in the methods of solution polymerization, melt polycondensation, interfacial polycondensation and solid polycondensation.

　　Unsaturated polyester is the product of polycondensation of unsaturated dicarboxylic acid or the product of saturated dicarboxylic acid and glycol reaction. In addition to containing ester groups, the polyester molecules also contain double bonds, and in the presence of the initiator, the polyester can react with the monomer to form thermosetting resin with cross-linked structure. Unsaturated polyester resin has low viscosity, good wetting properties, high transparency and a certain degree of adhesion. Polyester is used to make glass fiber reinforced plastics primarily, also used for clay, paint, and casting plastics.

　　Unsaturated polyester has many varieties, and the main difference among them is the raw materials and their proportions. We can get different toughness resin as well as a series of products such as flame retardant, electrical insulation and so on, to meet the requirements of various products. The common unsaturated acid is maleic anhydride, which is easy to obtain in industry. Common used diols include ethylene glycol, propylene glycol and diethylene glycol, etc. Azelaic acid, adipic acid and phthalic anhydride are common used saturated acids. These unsaturated acids can adjust the density of double bonds in the polyester chain, and aniline can also increase co-solubility of polyester and crosslinking agent styrene. Cross-linkers also include methyl methacrylate and diallyl phthalate, etc. The polymerization degree of crosslinking agent between crosslinking points is determined by the

实验六　不饱和聚酯树脂和玻璃纤维增强塑料的制备

一、实验目的
（1）了解控制线型聚酯聚合反应程度的原理及方法。
（2）掌握制备不饱和聚酯和玻璃纤维增强塑料的实验技能。

二、实验原理
　　逐步聚合是合成高分子材料的重要方法之一，如涤纶、尼龙、聚氨酯和醛树脂等常规高分子材料和聚碳酸酯、聚砜、聚苯醚、聚酰亚胺等高性能高分子材料都是通过逐步聚合制备的。逐步聚合是通过不同物质间官能团的化学反应而进行的，这些物质可以是单体，也可以是聚合度不同的分子链。逐步聚合的分子量随官能团反应程度的增高而逐渐增大，只有在很高的反应程度下才能生成高分子量的聚合物。按反应类型，逐步聚合反应可分为缩聚反应、逐步加聚反应和氧化偶联聚合等。按反应方式，逐步聚合反应可采用溶液聚合、熔融缩聚、界面缩聚和固相缩聚等。
　　不饱和聚酯是由不饱和二元酸、饱和二元酸和二元醇缩聚反应的产物。这类聚酯分子中除含有酯基外，还含有双键，在引发剂存在下，能与烯类单体进行共加聚反应，形成有交联结构的热固性树脂。不饱和聚酯树脂黏度低，浸润性好，透明度高，并且有一定的黏附力。主要用于制作玻璃纤维增强塑料，也可以用作胶泥、涂料及浇铸塑料等。
　　不饱和聚酯的品种有很多，主要是所用原料及配比不同，可以制得从刚性到韧性的树脂，以及具有阻燃性、电绝缘性等系列产品，广泛适应各种性能制品的要求。常用的不饱和酸是顺丁烯二酸酐，这是工业上易得的原料，常用的二元醇有乙二醇、丙二醇和一缩乙二醇等。常用的饱和酸有壬二酸、己二酸和苯酐等。这些饱和酸可以调节线型聚酯链中的双键密度，苯胺还能增加聚酯和交联剂苯乙烯的共溶性。交联剂还有甲基丙烯酸甲酯及邻苯二甲酸二烯丙酯等。交联点之间交联剂的聚合度大小，决定于不饱和聚酯中的双键与烯类单体的竞聚率及投料比。若聚酯中的双键密度大，交联点之间聚合度小，则交联密

reactivity ratio of double bond and olefin monomer in unsaturated polyester and feed ratio. If the density of double bond in polyester is high and the degree of polymerization between crosslinking points is small, the crosslinking density is high, which makes the polyester resin has low elasticity and good heat-resistance.

Glass fiber reinforced plastics have a lot of varieties, which is one of the development directions of modern plastics industry. The glass fiber is set as a filler normally and is known as the "glass steel". Unsaturated polyester and vinyl monomers are cross-linked in the presence of peroxide, and are coated with a glass cloth which has been pretreated, then get the polyester glass fiber. It can be used to make large parts of the aircraft, the hull, the train car, the building on the transparent corrugated board, chemical equipment and pipelines, etc. It has the characteristics of high tensile strength, light weight, good electric and heat insulation and so on.

Experimental study on glass fiber reinforced plastics is carried out in two steps. The first step is to make a linear unsaturated polyester by heating the melt condensation, and the reaction equation is shown in Figure 2.3.

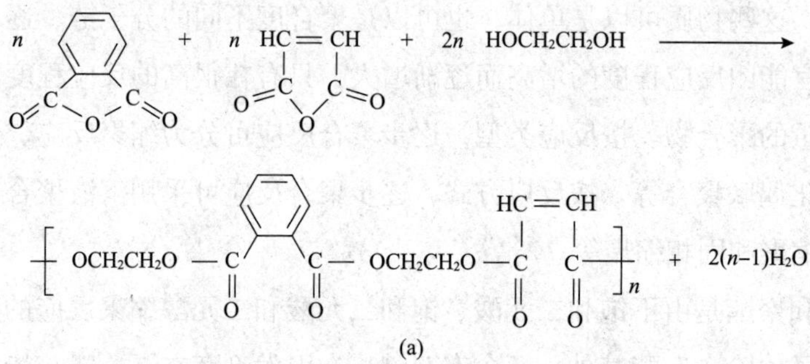

Figure 2.3 Synthesis of linear unsaturated polyester

In the process of polyester reaction, the degree of polymerization is controlled by frequent measurement of the acid value or dehydration amount. When the acid value is reduced to about 50, we can get a low viscosity liquid polyester (a). The polyester (a) and styrene containing inhibitor mixed to unsaturated polyester resin for standby. Styrene is both a diluent and cross-linking agent.

The second step is that linear-unsaturated resin is transfer into glass fiber reinforced plastics by crosslinking curing (Figure 2.4).

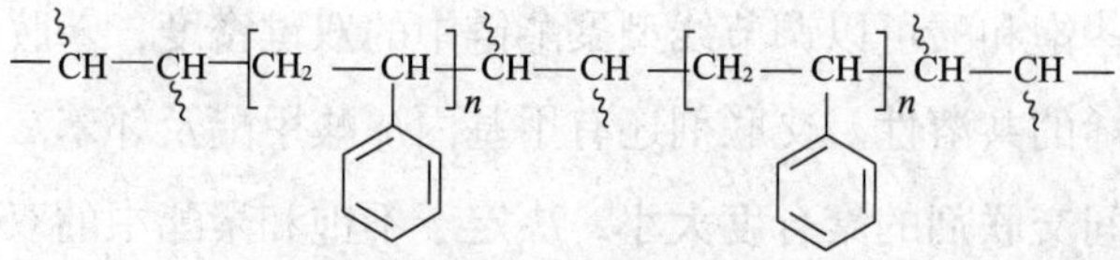

Figure 2.4 Crosslinking and curing reaction of linear polyester

度大，使聚酯树脂的弹性低，耐热性好。

玻璃纤维增强塑料品种有很多，是近代塑料工业发展方向之一。一般用玻璃纤维作填料，因此又称玻璃钢。聚酯玻璃钢是不饱和聚酯在过氧化物存在下与烯类单体交联之前，涂铺在经过预处理过的玻璃布上，在适当温度或低温低压下接触成型固化，可以用来制造飞机上的大型部件、船体、火车车厢、建筑上的透明瓦楞板、化工设备和管道等，具有抗张强度高、密度小、电和热的绝缘性优良等特点。

玻璃纤维增强塑料的实验分两步进行。第一步是由顺丁烯二酸酐、邻苯二甲酸酐和稍过量的乙二醇，通过加热熔融缩聚，制得线型不饱和聚酯，反应方程式如图 2.3 所示。

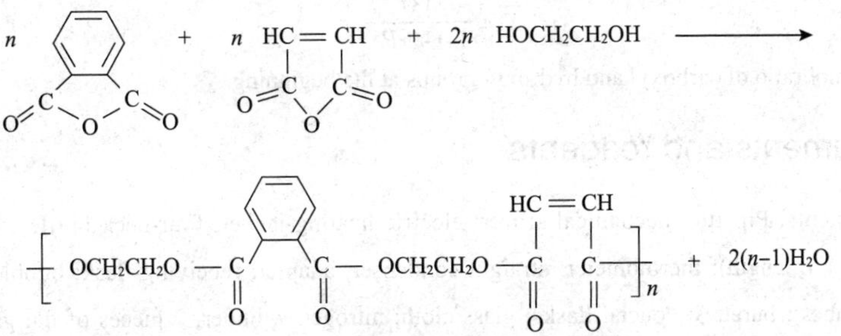

图 2.3 合成线型不饱和聚酯反应式

聚酯反应过程中，经常测定体系的酸值，或以脱水量来控制聚合度。当酸值降到 50 左右时，可以得到低黏度的液体聚酯（a）。将聚酯（a）和含有阻聚剂的苯乙烯混合制成不饱和聚酯树脂贮备待用。苯乙烯既是稀释剂又是交联剂。

第二步由线型的不饱和树脂交联固化成型制成玻璃纤维增强塑料，其交联固化反应式如图 2.4 所示。

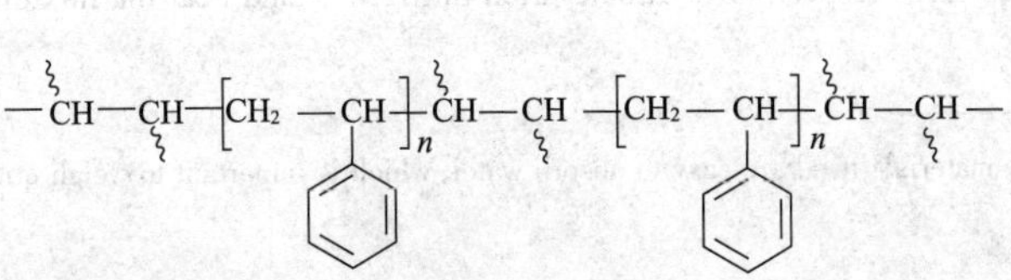

图 2.4 线型聚酯交联固化反应式

In the process of polymerization, acid value is used to evalute the degree of polymerization (DP). Acid value is defined as the amount (mg) of KOH required to neutralize the free acid contained in 1g of resin. Acid value A can be calculated by the following formula:

$$A = \frac{M(V-V_0) \times 56.1}{W} \times 1000 \tag{1}$$

where, M is the concentration of KOH (mol/L), W is the mass of sample (g), V is the consuming volume of KOH solution, V_0 is blank titration consumption volume.

If A_0 is the acid value of N_0 the carboxylic at the beginning of polymerization, when it comes to the time-t, there are only about N the carboxylic left with acid value A, and the degree of reaction is as followed:

$$P = \frac{N_0 - N}{N_0} = 1 - \frac{N}{N_0} = 1 - \frac{A}{A_0} \tag{2}$$

Average polymerization degree of the product:

$$\overline{X}_n = \frac{1+r}{1+r-2rP} \tag{3}$$

r- the mol ratio of carboxyl and hydroxyl groups at the beginning.

3. Instruments and reagents

Instruments: Pipette; mechanical stirrer; electric heating-jacket; four-neck bottles; fractionating column (10 cm length); thermometer; straight condenser; adapter; receiving flask; bubble counter (or U-shaped tubes); burettes; concial flasks; glass cloth; nitrogen cylinder; 2 pieces of flat glass (14cm × 12cm).

Reagents: Maleic anhydride; phthalic anhydride; glycol; styrene; hydroquinone; benzoyl peroxide; dibutyl phthalate; cobalt naphthenate; acetone; phenolphthalein- ethanol; KOH - ethanol; 0.2mol/L HCl.

4. Experimental procedure

(1) Synthesis of linear-unsaturated polyester

After drying, install the instrument according to Figure 2.5, and U-shaped tube is charged with an appropriate amount of paraffin oil. 24.5 g (0.25 mol) of maleic anhydride, 37.0 g (0.25 mol) of phthalic anhydride and 34.1 g (0.55 mol) of ethylene glycol ❶ are added to one four-neck round bottom bottle (250 mL), respectively. Bubble dried nitrogen ❷ and heat the mixture. After the reactants

---

❶ The raw materials used are easy to absorb water, which is important to weigh quickly so as not to affect the ingredients.

❷ In the early stage of reaction, do not pass nitrogen too fast, otherwise it will bring out the ethylene glycol and affect the ratio of raw materials.

在聚合过程中用酸值的大小来衡量聚合反应程度 P。酸值是指中和 1 g 树脂中所含的游离酸所需 KOH 的量（mg）。若 KOH 浓度为 M（mol/L），用它滴定 W g 样品，消耗了 V mL，空白滴定消耗 V_0 mL，则酸值按式（1）计算：

$$A = \frac{M(V-V_0) \times 56.1}{W} \times 1000 \tag{1}$$

若体系起始时 N_0 个羧基测得酸值为 A_0，聚合进行到时间 t 时，体系中残留 N 个羧基，此时酸值为 A，那么反应程度为式（2）：

$$P = \frac{N_0-N}{N_0} = 1 - \frac{N}{N_0} = 1 - \frac{A}{A_0} \tag{2}$$

根据反应程度 P，可求得产物平均聚合度式（3）：

$$\overline{X_n} = \frac{1+r}{1+r-2rP} \tag{3}$$

式中，r 是起始时羧基与羟基的摩尔比。

三、仪器和试剂

仪器：滴管；电动搅拌器；电热套；四口瓶；分馏柱（长 10 cm）；温度计；直形冷凝管；接应管；接收瓶；计泡器（或 U 形管）；滴定管；锥形瓶；氮气瓶；玻璃布；平板玻璃（14 cm×12 cm）2 块。

试剂：顺丁烯二酸酐；邻苯二甲酸酐；乙二醇；苯乙烯；对苯二酚；过氧化苯甲酰；邻苯二甲酸二丁酯；环烷酸钴；丙酮；酚酞－乙醇溶液；KOH－乙醇溶液；0.2 mol/L HCl。

四、实验步骤

（1）合成线型不饱和聚酯

干燥后安装好的仪器如图 2.5 所示，U 形管中加入适量的石蜡油。称 24.5 g（0.25 mol）的顺丁烯二酸酐，37.0 g（0.25 mol）邻苯二甲酸酐和 34.1 g（0.55 mol）的乙二醇❶先后加入 250 mL 的四口瓶内。通干燥氮气❷，加热，待反应物熔融

---

❶ 所用原料都极易吸水，称重要迅速，以免影响配料比。

❷ 反应前期，通氮气不可过快，否则会带出乙二醇，影响原料配比。

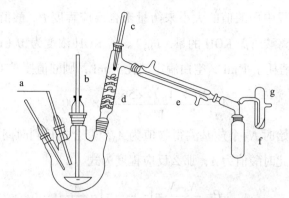

Figure 2.5 Apparatus preparation of unsaturated polyester
a– nitrogen catheter, b– stirrer, c– thermometer, d– fractionating column, e– condenser, f– receiving flask, g– bubble counter

are melted, start the agitator, and the temperature rises sharply. Slow down the heating rate ❶ and gradually increase to 160℃ within about 1 h. The esterification reaction starts when water appears on the glass wall, keeping reacting for 1.5 h at 160℃, then heat to 190-200℃ with the speed of nitrogen becomes slightly faster, maintaining at 200℃ for 1.5h, finally determine the acid values (determination methods, see in Note ❷). Sample at every 30 minutes until the acid value dropped to about 50, and stop heating. The polymerization reaction lasted for about 6 hours to obtain a transparent and yellowish viscous liquid. Take advantage of the heat to open the four-neck bottle and pour out the viscous liquid timely.

(2) Preparation of the unsaturated polyester - styrene solution

Add 0.01 g ❸ of hydroquinone and 20 g of styrene into 100mL beaker, weigh together with the beaker to get (W_1), then put the polyester into styrene beaker while stirring with a glass rod. Weigh beaker

---

❶ At the initial stage of the reaction (at 140℃), the temperature of the liquid will automatically increase due to the exothermic reaction, so it will be heat up slowly to avoid burst out of product. When the monomer is gradually transformed into an oligomer, the temperature can be raised to about 190℃ (the boiling point of ethylene glycol is 197.2℃). In the middle and later stage, it is important to control the reaction temperature. However, the high temperature can produce side effects, which affect the quality of the resin.

❷ Determination of resin acid: 20 mL of acetone is poured in 5 250mL-conical flasks. Stuff bottle with a cork, then weigh accurate and label it. Use a long dropper to suck about 1g of resin into a conical flask containing acetone, plug it and weigh accurately, shake the conical flask to dissolve the resin completely, then add three drops of phenolphthalein ethanol solution, titrate with 0.2mol/L KOH-ethanol standard solution until the light red color does not fade. And do a blank experiment.

❸ Maleic acid is not easy to self polymerize, but when maleic acid is made of unsaturated polyester and styrene, it is easy to crosslink, so the preparation of unsaturated polyester-styrene solution should be added some inhibitors.

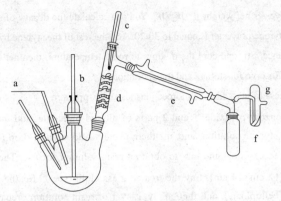

图 2.5 制备不饱和聚酯装置图
a—氮气导管，b—搅拌，c—温度计，d—分馏柱，e—冷凝管，f—接收瓶，g—计泡器

后，开动搅拌器，此时温度急剧上升，待液温升至130℃后，减慢升温速度❶，约 1 h 内逐步升至 160 ℃。当玻璃壁上出现水珠时说明酯化已开始，保持在 160℃反应 1.5 h 后，再升温至 190~200℃，通氮气速度稍加快，保持在 200℃ 反应 1.5h 后，取样测定酸值❷。以后每隔 30 min 取样一次，直至酸值降至 50 左右时，停止加热，聚合反应共 6 h 左右，得到透明略带黄色的黏稠液。趁热打开四口瓶倒出黏稠液。

（2）不饱和聚酯—苯乙烯溶液的配备

先称 20 g 的苯乙烯放入 100 mL 烧杯中，并加入 0.01 g 的对苯二酚❸，连同烧杯称重（W_1），然后将自行冷却至 90℃的聚酯倒入盛有苯乙烯的烧杯中，立即用玻璃棒搅拌均匀，并连同烧杯称重（W_2），则得到聚酯净重 $W=W_2-W_1$。再按苯乙烯/聚酯 =30/70 质量比算出苯乙烯用量，除已加的 20 g 外，按计量补加

---

❶ 反应初期（140℃左右），由于放热反应，液温会自动上升，因此要减速升温，以免引起冲料。待单体逐步转变成低聚物后，才能升温至190℃左右（乙二醇的沸点 197.2℃）。反应中期和后期也要控制好温度，高温有利于酯化反应，但过高的温度会产生副反应，影响树脂质量。

❷ 树脂酸值的确定。预先在 5 只 250 mL 的锥形瓶中分别加 20 mL 丙酮，塞好塞子后准确称重，贴上标签，待用。用长滴管吸取1g左右的树脂滴入一只盛有丙酮的锥形瓶中立即塞好，准确称重，摇动锥形瓶使树脂全部溶解，然后加入三滴酚酞 – 乙醇溶液，用 0.2 mol/L 的 KOH– 乙醇标准溶液滴定至淡红色不褪为终点。并做一个空白试验。

❸ 顺丁烯二酸酐于不易自聚，但是当制成不饱和聚酯后与苯乙烯很容易共聚交联，因此配制不饱和聚酯 – 苯乙烯溶液时应加阻聚剂。

then tend to obtain polyester net weight $W=W_2-W_1$. You can calculate the dosage of styrene according to that the mass ratio of styrene/polyester is equal to 30/70. Add the rest of the styrene by calculation besides the 20 g added in advance. Stir and cool the mixture to room temperature, then get the transparent light yellow viscous liquid. Reserve for storage and transportation.

(3) Synthesis process of glass fiber reinforced plastics at low pressure

Mix 2 parts of benzoyl peroxide ❶ and 2 parts of dibutyl phthalate, add them to 100 parts of unsaturated polyester styrene solution, and stir them evenly. Add a small drop (about 0.01g) of N, $N-$ dimethylaniline (accelerator) and mix to obtain a resin solution for use. The silicone is coated on clean plate glass (12 cm×14 cm). Lay the treated glass cloth (see ❷ for the treatment method) on the glass plate (or cellophane), and then apply a layer of resin solution prepared above to make the glass cloth soak. Use a brush or a thick glass rod to extrude the bubbles between the resin and the glass cloth, and then apply it repeatedly until the required thickness (6-8 layers). Then cover with cellophane, finally press on the glass plate (four corners should be coated with silicone oil), clamp with four clamps and wipe the resin on the edge. Cure it and moving into the oven at 80 ℃ for 0.5h, then remove the glass plate with a screwdriver and cut off the edge, move into the oven again at 80 ℃ for 1.5h. Cool to indoor temperature, stripping, which is, the glass fiber reinforced plastics.

Questions

(1) If you want to prepare a good toughness, flexible glass steel, how to design the ratio of ingredients?

(2) The molar coefficient ($r\leqslant 1$) can be calculated according to the feed ratio of raw materials. The polymerization degree p and average polymerization degree Xn can be calculated according to the final acid value (Tip: calculate the initial acid value first).

---

❶ The initiator and promoter do not mix directly. Resin must be cooled to room temperature, then add initiator complexing agent. If you want to produce large devices, it only can be cured at room temperature usually by adding accelerator like cobalt naphthenate ($(C_{11}H_7COO)_2Co$), etc, it can react at 25℃ for 6 h if use 0.25% oxidation butanone and 0.02% cobalt naphthenate.

❷ Treatment of glass cloth: The advantages and disadvantages of glass fiber reinforced plastics depend on the nature of the resin, glass fiber strength and the adhesion between the resin and glass fiber. Glass cloth processing methods include heat treatment and water washing. In order to improve the surface properties of glass fiber and enhance the ability of adhesion between glass fiber and resin, sometimes it is necessary to remove the surface protecting agent using surface treament agen. In this experiment, the glass cloth can be immersed in 20% soap solution and boiled for 20 minutes, then washed with water and dried for further use.

入剩余的苯乙烯，搅拌均匀，冷却至室温，即得黏稠的、透明带淡黄色的黏稠液。可以贮藏和运输。

（3）玻璃纤维增强塑料的低压成型

将过氧化苯甲酰2份❶，邻苯二甲酸二丁酯2份混合均匀后，加到不饱和聚酯-苯乙烯溶液100份中，搅拌均匀，再用滴管加一小滴（约0.01g）的 N,N—二甲基苯胺（促进剂），混合均匀即得树脂溶液，立即使用。在两块清洁的平板玻璃（12cm×14cm）上涂少量硅油（或铺一层玻璃纸），将处理过的玻璃布❷，铺在玻璃板或玻璃纸上，然后涂上一层配制好的树脂溶液，使玻璃布浸润，用刷子或粗玻棒挤出树脂和玻璃布之间的气泡，这样反复涂铺，直至需要的厚度（6~8层），然后盖上玻璃纸，最后压上玻璃板（四个角要涂上硅油），用4只夹子夹紧，擦净边缘的树脂，平放在室温下待初步固化后，移入80℃烘箱固化0.5h，取出后用起子去掉玻璃板，用剪刀剪去边缘部分，再移入80℃烘箱进一步固化1.5h。冷却至室温，脱模，即得玻璃纤维增强塑料。

思考题

（1）若要制备韧性好、柔性大的玻璃钢，应如何设计配料？

（2）按原料的投料比计算摩尔系数（$r \leqslant 1$），按最后酸值计算聚合反应程度 P 及平均聚合度 X_n（提示：首先计算起始的酸值）。

---

❶ 引发剂不要直接与促进剂混合。树脂必须冷却至室温后，才能加入引发剂等配合剂。若制备大型器件，只能室温固化，一般加入环烷酸钴（$(C_{11}H_7COO)_2Co$）等促进剂，如用0.25%的过氧化丁酮及0.02%的环烷酸钴，在温度为25℃下固化6h即可。

❷ 玻璃布的处理。玻璃纤维增强塑料的优劣，取决于树脂的性质、玻璃纤维的强度以及树脂与玻璃纤维之间的粘接力等方面。玻璃布的处理方法有热处理和水洗法，除去其表面保护剂，有时需要表面处理剂，改善玻璃纤维表面性能，增强玻纤与树脂之间的粘接能力。本实验用水洗法，即将玻璃布浸入20%肥皂液中煮洗20min，然后用水冲洗干净，烘干备用。

Experiment 7　Solution polymerization of acrylamide

1. The purposes

(1) Master the methods and principle of solution polymerization.

(2) Learn how to select the solvent.

2. Experimental principle

Solution polymerization is a kind of polymerization methods in which monomers are dissolved in the solvent to polymerize. The solution polymerization system is mainly composed of monomer, initiator or catalyst and solvent. Some of the resulting polymers are dissolvable in solvent, and others are insoluble. The synthetic method of the former is called as homogeneous polymerization, and the latter is precipitation polymerization. Solution polymerization method is available for radical polymerization, ionic polymerization and polycondensation.

In precipitation polymerization, polymers are in a poor solvent, whose chains are in the crimp and end groups are wrapped, thus automatic acceleration occurs in the initial stage of polymerization. There is no steady-state phase in precipitation polymerization. With monomer conversion increasing, aggragation degree deepens and automatic acceleration effect increases correspondingly. The kinetic behavior of precipitation polymerization is obviously different from that of homogeneous polymerization. In homogeneous polymerization, the polymerization rate is proportional to the square root of the initiator concentration according to the biradical termination mechanism. However, precipitation polymerization is unstable at the beginning. With the deepening of the encapsulation degree, the reaction can only be terminated by single radical. Therefore, the polymerization rate will be proportional to the first power of initiator concentration.

In homogeneous solution polymerization, polymers are in a good solvent, whose chains are in the stretch state. The extent of aggragation is shallow, and chain segments diffuse easily. Thus active end groups approach each other, and biradical termination occurs easily. Only at high conversion rates, automatic acceleration begins to appear. If the monomer concentration is low, it is possible to eliminate the automatic acceleration effect, so the reaction follows normal radical polymerization kinetics. Thus solution polymerization is one of the methods that commonly used in the laboratory to study polymerization mechanism and kinetics.

The solvent used in solution polymerization is not entirely inert. It has various effects on the reaction. The following issues should be considered when selecting solvents:

(1) The influence on initiator's decomposition. The decomposition rate of azo initiators (e.g. azobisisobutyronitrile) is less affected by solvent, but solvent has a greater induce decomposition effect

实验七　丙烯酰胺水溶液聚合

一、实验目的

（1）掌握溶液聚合的方法和原理。
（2）学习如何选择溶剂。

二、实验原理

单体溶于溶剂中而进行的聚合方法叫作溶液聚合。溶液聚合体系主要由单体、引发剂或催化剂溶剂组成。生成的聚合物有的溶解，有的不溶，前一种叫作均相聚合，后一种叫作沉淀聚合。自由基聚合、离子聚合和缩聚均可用溶液聚合的方法。

在沉淀聚合中，由于聚合物处在非良溶剂中，聚合物链处于卷曲状态，端基被包裹，聚合开始就出现自动加速现象，不存在稳态阶段。随着转化率的提高，包裹程度加深，自动加速效应相应增强。沉淀聚合的动力学行为与均相聚合有明显不同。均相聚合时，依双基终止机理，聚合速率与引发剂浓度的平方根成正比。而沉淀聚合一开始就是非稳态，随包裹程度的加深，只有单基终止反应途径，故聚合速率将与引发剂浓度的一次方成正比。

在均相溶液聚合中，由于聚合物是处在良溶剂环境中，聚合物链处于比较伸展的状态，包裹程度浅，链段扩散容易，活性端基容易相互靠近而发生双基终止。只有在高转化率时，才开始出现自动加速现象。若单体浓度不高，则可能消除自动加速效应，使反应遵循正常的自由基聚合动力学规律。因而溶液聚合是实验室中研究聚合机理及聚合动力学等常用的方法之一。

进行溶液聚合时，由于溶剂并非完全是惰性的，其对反应会产生各种影响。选择溶剂时应考虑到以下几个问题。

（1）对引发剂分解的影响。偶氮类引发剂（如偶氮二异丁腈）的分解速率受溶剂的影响很小，但溶剂对有机过氧化物引发剂有较大诱导分解作用。这种作用按下列顺序依次增大：芳烃＜烷烃＜醇类＜醚类＜胺类。诱导分解的结果

for the organic peroxide initiators. This effect increases successively in the following order: aromatic < hydrocarbons < alkanes < alcohols < ethers < amines. Induce decomposition of initiators results in low initiation efficiency.

(2) The chain transfer effect of solvent. Free radical is a very active reaction center, which not only can initiate monomer molecules, but also can react with solvent to capture an atom, such as hydrogen or chlorine, to meet its unsaturated valence. The stronger the ability of solvent molecules to provide such atoms, the stronger the chain transfer. As a result of chain transfer, the molecular weight of the polymer decreases. If the activity of the free radicals generated by the reaction decreases, the polymerization rate will also decrease.

(3) The effect on the solubility of polymer. Solubility properties of Solvent for polymer control the morphology (twist or stretch) and viscosity of active chains, which determines chain termination speed and molecular weight distribution.

Compared with the bulk polymerization, solution polymerization has many advantages, such as the lower viscosity, easier mixing and heat transfer, avoid local overheating, and easy control temperature, etc. However, solution polymerization is rarely used in industry due to the high cost of organic solvents and difficulty in recovery. It only used in the case of using directly polymer solution, such as coatings, adhesives, impregnating agents and synthetic fibers, etc.

Acrylamide is a kind of water-soluble monomer and polyacrylamide is also soluble in water. This experiment selects water as the solvent in solution polymerization, which has the advantages of low cost, non-toxic, small chain transfer constant, good solubility for monomer and polymer. It belongs to homogeneous polymerization.

As an excellent flocculant, polyacrylamide has good water-soluble property and is widely used in oil exploration, mineral processing, chemical industry and sewage treatment, ect.

3. Instruments and reagents

Instruments: three-neck round bottom flask (250 mL); spherical condenser; thermometer; mechanical stirrer; Y-shaped tube; electronic balance; beaker (500 mL); cylinder (100 mL, 10 mL); vacuum filtration device; glass rod; watch glass.

Reagents: acrylamide; methanol; ammonium persulfate; distilled water.

4. Experimental procedure

The middle port of three-neck flask (250 mL) is fitted with a mechanical stirrer. A thermometer and the condenser are mounted on the other ports. Add 10 g (0.14 mol) of acrylamide and 80 mL of distilled water into the reaction flask, start to stir and heat to 30 ℃ ❶ by the water bath (see notes) for dissolving

---

❶ In the case of exhausting the oxygen, we can obtain polymer with high molecular weight. If using low-temperature reverse phase emulsion polymerization, it is possible to obtain the higher molecular weight's polymer.

使引发剂的引发效率降低。

（2）溶剂的链转移作用。自由基是一个非常活泼的反应中心，它不仅能引发单体分子，而且还能与溶剂反应，夺取溶剂分子中的一个原子，如氢、氯，以满足其不饱和原子价。溶剂分子提供这种原子的能力越强，链转移作用就越强。链转移的结果使聚合物分子量降低。若反应生成的自由基活性降低，则聚合速度也将减小。

（3）对聚合物的溶解性能影响。溶剂溶解聚合物的性能控制着活性链的形态（蜷曲或舒展）及其黏度，它们决定了链终止速度与分子量的分布。

与本体聚合相比，溶液聚合体系具有黏度较低、混合及传热较容易、不易产生局部过热、温度容易控制等优点。但由于有机溶剂费用高、回收困难等原因，使得溶液聚合在工业上很少应用，只在直接使用聚合物溶液的情况，如涂料、胶黏剂、浸渍剂和合成纤维纺丝液等采用溶液聚合的方法。

丙烯酰胺为水溶性单体，其聚合物也溶于水。本实验采用水为溶剂进行溶液聚合，其优点是价廉、无毒、链转移常数小、对单体及聚合物溶解性能都好，为均相聚合。

聚丙烯酰胺是一种优良的絮凝剂，水溶性好，被广泛应用于石油开采、选矿、化学工业及污水处理等方面。

三、仪器和试剂

仪器：三口烧瓶（250 mL）；球形冷凝管；温度计；机械搅拌器；Y形管；电子天平；烧杯（500 mL）；量筒（100 mL、10 mL）；减压抽滤装置；玻璃棒；表面皿。

试剂：丙烯酰胺；甲醇；过硫酸铵；蒸馏水。

四、实验步骤

在 250 mL 三口反应瓶的中间口装上搅拌器，在一个侧口装上温度计，另一侧口装上冷凝管。将 10 g（0.14mol）丙烯酰胺和 80 mL 蒸馏水加入反应瓶中，开动搅拌，用水浴加热至 30℃ ❶，使单体溶解。然后把溶解于 10 mL

---

❶ 在排除氧气的情况下，可以获得分子量比较高的聚合物。若采用低温逆相乳液聚合，则可以获得分子量很高的聚合物。

monomers. Then add 0.05 g ammonium persulfate dissolved in 10 mL distilled water into the reaction flask from the upper port of the condenser tube, and flush the condenser tube with another 10 mL distilled water. The temperature was gradually raised to 90 ℃ (observe "climbing rod" phenomenon and explain it) and the polymer began to form. The reaction temperature was 90 ℃ for 2-3 h. After the reaction, the product was poured into a 500 mL beaker containing 150 mL methanol, and the polyacrylamide precipitated at the same time. After standing for a while, add a small amount of methanol into the beaker and observe whether there is still precipitation. If there is any, a small amount of methanol can be added to make the polymer precipitate completely, and then use the Buchner funnel to filter. After washing three times with a small amount of methanol, the precipitate was transferred to a watch glass and dried to constant weight in a vacuum oven at 30 ℃. Weigh and calculate the yield.

Questions

(1) When choosing the solvent, what problems should be paid attention to in solution polymerization?

(2) In what situation do we can adopt solution polymerization in the industry?

(3) Why should the temperature gradually rise during the polymerization?

Experiment 8　Synthesis of super absorbent material —— low cross-linking sodium polyacrylate

1. The purposes

(1) Synthesize sodium polyacrylate with ultra-absorbent property.

(2) Understand the principle and method of inverse suspension polymerization.

2. Experimental principle

Water absorbing capacity of traditional absorbent materials such as paper, cotton, foam plastic, etc, is the 10-20 times of themselves' weight. In recent years, a series of materials with ultra-high water absorbing capacity have begun to emerge, and their water absorption capacity is up to hundreds or even thousands times of their own weight. These materials can be used as industrial dehydrators and thickeners, sanitary materials such as diapers, and soil conservation in agriculture and horticulture. Because of the broad application prospects of these materials, they are paid more and more attention.

Ultra-absorbent materials is usually synthesized by slightly crosslinking of some water-soluble

蒸馏水中的0.05 g过硫酸铵从冷凝管上口加入到反应瓶中，并用10 mL蒸馏水冲洗冷凝管。逐步升温至90℃（观察爬杆现象，并解释），开始生成聚合物。在90℃下反应2~3h。反应完毕后，将所得产物倒入盛有150 mL甲醇的500 mL烧杯中，边倒边搅拌，这时聚丙烯酰胺便沉淀出来。静置片刻，向烧杯中加入少量甲醇，观察是否仍有沉淀生成。若还有则可再加少量甲醇，使聚合物沉淀完全，然后用布氏漏斗抽滤。沉淀用少量甲醇洗涤三次后，转移到表面皿上，在30℃真空烘箱中干燥至恒重。称重，计算产率。

思考题

（1）进行溶液聚合时，选择溶剂应注意哪些问题？

（2）工业上在什么情况下采用溶液聚合？

（3）聚合反应中为什么要逐步升温？

实验八　超高吸水性材料——低交联聚丙烯酸钠的合成

一、实验目的

（1）合成一种具有超高吸水性能的聚丙烯酸钠。

（2）了解反相悬浮聚合的原理和方法。

二、实验原理

传统的吸水性材料，如纸、棉、泡沫塑料等，只能吸收自重的10~20倍的水。近年来，人们合成了一系列具有超高吸水性能的材料，其吸水量可达自重的几百倍乃至上千倍。这些材料可以作为工业用脱水剂和增稠剂，用作纸尿片等卫生材料，以及用于农业和园艺栽培上的土壤保持。由于具有广阔的应用前景，这类材料越来越受到人们的重视。

制备超高吸水性材料，通常是将一些水溶性高分子，如聚丙烯酸、聚乙

polymer, such as poly (acrylic acid), poly (vinyl alcohol), polyacrylamide, poly (ethylene oxide), poly (vinyl pyrrolidone), etc.

The property and amount of crosslinking agent have a great influence on the water absorption capacity of the final product. When the amount of crosslinking agent is small, part of the polymer will be soluble in water, however the excessive amount of crosslinking agent lead to high cross-linking degree and low swelling degree, both of which can decrease water absorbing ability.

For sodium polyacrylate prepared by methyl bisacrylamide crosslinking, the degree of neutralization of poly (acrylic acid) also has a significant impact on water absorption.

In order to control the reaction conditions and simplify purification treatment of the final product, such polymerizations are particularly suitable for the inverse suspension or inverse emulsion polymerization systems, that is, polymerization starts with concentrated aqueous monomers. Generally, the aqueous solution containing 50%-80% monomers and water-soluble initiators (e.g., persulfate, hydrogen peroxide and redox initiator) are dispersed together in aliphatic hydrocarbon as the continuous phase. Suspension stabilizers contain sorbitol oil esters, soluble polymers having a -COOH, -SO_3H and –NH_2 and other hydrophilic substituents.

In order to increase the absorption rate of final product, it is advantageous to add a small amount of surfactants before drying the polymer.

3. Instruments and reagents

Instruments: three-neck flask (250 mL); water bath; beaker (100 mL × 2, 1000 mL); mechanical stirrer; condenser; glass stopper; cylinder (25 mL × 2, 100 mL); 3 burettes; glass rod; ice water bath; electronic balance; tray balance; cloth and thread for water absorption measurement; oven.

Reagents: acrylic acid; 16% NaOH; N,N-methylene bisacrylamide; Span-60; OP emulsifier; $K_2S_2O_8$; n-hexane.

4. Experimental procedure

(1) A 100 mL beaker containing 10 mL of acrylic acid was placed in an ice water bath for cooling. 16% NaOH solution (20 mL) was slowly dropped to the beaker under stirring. Then the mixture was transferred to a 250 mL three-neck reaction flask equipped with a reflux condenser tube, mechanical stirrer and a capillary tube for purging N_2. The emulsifier span-60 (0.6 g), crosslinker N, N-methylene bisacrylamide (BMA) (0.006 g), initiator $K_2S_2O_8$ (0.047 g) was added successively, and then add 44 mL of n-hexane under stirring.

(2) Heat reaction flask in the water bath, and start stirring ❶. When the temperature rises to 65 ℃, maintain the reaction at 65 ℃ for 3h. Stopping the reaction and pour the reaction mixture into a 150 mL

---

❶ To get stable suspension system, the reaction should be under a good stirring, otherwise polymers will agglomerate in the bottom of bottle.

烯醇、聚丙烯酰胺、聚氧化乙烯、聚乙烯基吡咯烷酮等进行轻微的交联而得到。

交联剂的性质和用量，对最终产物的吸水能力影响很大。用量少时，部分聚合物会溶于水，而用量过多则使交联度过大，溶胀度降低，两者都会使吸水能力降低。

对于用 N,N- 亚甲基双丙烯酰胺交联的聚丙烯酸钠来说，聚丙烯酸的中和程度对吸水量也有很大影响。

为便于控制反应条件和简化对最终产物的后处理，这类反应特别适合在逆悬浮或逆乳液聚合体系中进行，即聚合是从浓的单体水溶液开始的。一般是将含 50%~80% 单体的水溶液与水溶性引发剂（如过硫酸盐、过氧化氢和氧化还原引发剂）一起分散，常用脂肪烃作为外分散相。悬浮稳定剂有山梨糖醇脂肪酸酯和带有—COOH，—SO_3H 及—NH_2 等亲水性取代基的可溶性聚合物。

为了提高最终产物的吸水速率，还常常在聚合物干燥前加入少量的表面活性剂。

三、仪器和试剂

仪器：三口反应瓶（250 mL）；水浴锅；烧杯（100 mL×2，1000 mL）；搅拌器；冷凝管；玻璃塞；量筒（25 mL×2，100 mL）；滴管（3 支）；玻璃棒；冰水浴；电子天平；托盘天平；吸水率测量用布和线；烘箱。

试剂：丙烯酸；16% NaOH；N,N- 亚甲基双丙烯酰胺（MBA）；司班 -60（Span-60）；OP 乳化剂；$K_2S_2O_8$；正己烷。

四、实验步骤

（1）将盛有 10 mL 丙烯酸的 100 mL 小烧杯置于冰水浴中冷却，在搅拌下慢慢滴入 16% NaOH 溶液 20 mL，然后将混合液转移到带有回流冷凝管、搅拌器及通 N_2 毛细管的 250 mL 三口反应瓶中，依次加入乳化剂司班 -60 0.6 g，N,N- 亚甲基双丙烯酰胺 0.006 g，引发剂 $K_2S_2O_8$ 0.047 g，再在搅拌下加入正己烷 44 mL。

（2）用水浴加热反应瓶，并开动搅拌❶，待温度升至 65℃，维持在该温度

---

❶ 为能得到稳定的悬浮体系，应保持反应体系有良好的搅拌，否则聚合物会在瓶底结块。

beaker. Then decant the upper layer of n-hexane (filter if necessary) and obtaine the hydrogel.

(3) Add 0.33 mL of OP emulsifier to the polymer particles, mix them thoroughly, and dry them in the oven at 50℃. Note that baking temperature is not too high, otherwise it will produce yellow and parched product.

(4) Weigh 0.1g of the above powder sample into a 100 mL beaker, then add 60~70 mL distilled water and swell the powder for 4-5 days ❶ under gently stirring. Pour the gel and water mixture on the weighed nylon gauze and filter it (tie the nylon gauze on the beaker with rubber band). After 15 minutes of natural dripping filtration, it is weighed together with the filter cloth. The weight of the absorbent material divided by the weight of the dry powder sample is regarded as the water absorption capacity of the polymer, which is usually expressed by g pure water / g polymer. The water absorption of the product was calculated.

Questions

(1) Compare the similarities and differences between the inverse suspension polymerization and general suspension polymerization, and illustrate their respective characteristics.

(2) Analyse the main factors that influence the water absorption of final polymerization product.

Experiment 9   Synthesis of polyvinyl acetate latex

1. The purposes

(1) Understand the characteristics, recipe and function of each component in emulsion polymerization.

(2) Be familiar with the sunthesis process and application of poly (vinyl acetate) latex.

2. Experimental principle

Emulsion polymerization refers to the heterogeneous polymerization of monomers dispersed in the medium with the help of emulsifiers and water-soluble initiators, which occurs under stirring or vivration. It is different from the solution polymerization and suspension polymerization. Emulsifier is the main component of emulsion polymerization. It is composed of non-polar hydrophobic group and polar hydrophilic group. The hydrophile lipophilic balance value (HLB value) is usually used to

---

❶ Both swelling time and natural trickling filtration time have obvious influence on the measured value of water absorption of polymer. If time permits, the swelling time can be extended, such as 12 h.

反应 3 h。停止反应，将反应混合物倒入 150 mL 烧杯中，倾出上层正己烷（必要时进行过滤），分离出水凝胶部分。

（3）向聚合物胶粒中加 0.33 mL OP 乳化剂，充分混合后，在烘箱中 50℃ 烘干。注意烘烤温度不宜太高，否则将导致产物变黄、变焦。

（4）称取上述粉末样品 0.1g，置入 100 mL 烧杯中，加入蒸馏水 60~70mL，溶胀 4~5 天❶，同时轻轻搅动。然后倒在已称重过的尼龙纱布上过滤（用橡皮筋将尼龙纱布扎在大烧杯口上）。让其自然滴滤 15 min 后，连同滤布一起称重，以吸水后聚合物的质量除以干粉末样品的质量，视为该聚合物的吸水能力，通常用纯水（g）/聚合物（g）来表示。计算产物的吸水率。

思考题

（1）试比较逆悬浮聚合和普通悬浮聚合的异同和各自的特点。

（2）分析影响最终聚合产物的吸水力的主要因素。

实验九　聚醋酸乙烯酯胶乳的制备

一、实验目的

（1）了解乳液聚合特点、配方及组分的作用。

（2）熟悉聚醋酸乙烯酯胶乳的制备及用途。

二、实验原理

乳液聚合是指单体在乳化剂的作用下分散在介质中，加水溶性引发剂，在搅拌或震荡下进行的非均相聚合反应。它既不同于溶液聚合，也不同于悬浮聚合。乳化剂是乳液聚合的主要成分，它由非极性的亲油基和极性的亲水基组成，为了表示乳化剂的亲水性或亲油性，通常采用亲水亲油平衡值（HLB 值）

---

❶　溶胀时间和自然滴滤时间的长短都对聚合物吸水率的测定值有明显影响。若时间允许可延长溶胀时间，比如 12 h。

indicate the hydrophile or lipophilicity of emulsifier. That is, the lower the HLB value is, the stronger its hydrophobicity is; the higher the HLB value, the stronger hydrophilicity. According to the properties of polar groups, emulsifiers can be divided into anionic, cationic and nonionic types. The initiation, propagation and termination of emulsion polymerization are carried out in the micelles latex particles. Monomer droplets only functions as a reservoir. The polymerization rate is mainly determined by the number of particles, which has the characteristics of fast speed and can obtain high molecular weight products.

Emulsion polymerization mechanism of vinyl acetate latex is the same as the general one. Selecting persulfate as initiator, monomer and initiator are added in batches to make the reaction proceed smoothly. The common emulsifier in polymerization is poly (vinyl alcohol). In practice, the emulsifying effect and stability of the two emulsifiers are better than that of the single emulsifier. In this experiment, poly (vinyl alcohol) and OP-10 were used as combinational emulsifiers.

Poly (vinyl acetate) latex paint has the advantages of water-based paint, which has a small viscosity, a large molecular weight, and is not flammable. When used as adhesive (commonly known as white glue), it can be used in wood, fabric and paper.

3. Instruments and reagents

Instruments: Four-neck round bottom bottle (250 mL); mechanical stirrer; a transformer; super thermostat bath; a dropping funnel; spherical condenser tube; a thermometer; cylinders (100 mL, 50 mL, 10 mL); beakers (100 mL, 50 mL, 10 mL); a glass rod.

Reagents: Vinyl acetate; ammonium persulfate; poly (vinyl alcohol); emulsifier OP-10; dibutyl phthalate; sodium bicarbonate.

4. Experimental procedure

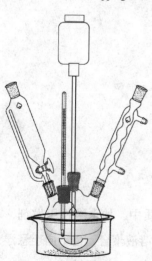

Figure 2.6　Schematic diagram of reaction device

As shown in Figure 2.6, add emulsifier (3 g poly(vinyl alcohol) and 0.5 g OP-10 dissolved in 40mL distilled water at 90 ℃ for about 1-2 h) and 10.7 mL of vinyl acetate in the four-neck round bottom bottle equipped with agitator, reflux condensation tube, dropping funnel and thermometer. After all emulsifiers are dissolved, weigh 0.5 g of ammonium persulfate in 2.5 mL of distilled water to make initiator solution. Half of the solution is poured into the reactor under stirring ❶. Heat the constant temperature bath to 65 ~ 70 ℃ and add 16 mL of vinyl acetate (dropping speed should not be too

---

❶ Keep a constant stirring speed during the polymerization, otherwise the emulsification of vinyl acetate would be incomplete.

来表示，HLB 值越低，其亲油性越强；反之，HLB 值越高，其亲水性越强。根据极性基团的性质可将乳化剂分为阴离子型、阳离子型和非离子型。乳液聚合的引发、增长和终止都在胶束的乳胶粒内进行。单体液滴只是储藏单体的仓库。反应速率主要决定于粒子数，具有快速、分子量高的特点。分子中有亲水基和亲油基。各种乳化剂的 HLB 值不同，为了获得稳定的乳状液，必须选择合适的乳化剂。

醋酸乙烯酯胶乳聚合机理与一般乳液聚合相同。采用过硫酸盐为引发剂，为使反应平稳进行，单体和引发剂均需分批加入，聚合中常用的乳化剂是聚乙烯醇。实践中还常把两种乳化剂合并使用，乳化效果和稳定性比单独用一种好。本实验采用聚乙烯醇和 OP-10 两种乳化剂。

聚醋酸乙烯酯胶乳漆具有水基涂料的优点，黏度小，分子量较大，不用易燃的有机溶剂。作为胶黏剂时（俗称白胶），木材、织物和纸张均可使用。

三、仪器和试剂

仪器：四口烧瓶（250 mL）；机械搅拌器；超级恒温槽；滴液漏斗；球形冷凝管；温度计；量筒（100 mL、50 mL、10 mL 各 1 支）；烧杯（100 mL、50 mL、10 mL 各 1 只）；玻璃棒。

试剂：醋酸乙烯酯；过硫酸铵；聚乙烯醇；乳化剂 OP-10；邻苯二甲酸二丁酯；碳酸氢钠。

四、实验步骤

如图 2.6 所示，在装有搅拌器、回流冷凝管、滴液漏斗及温度计的四口烧瓶中加入乳化剂（3 g 聚乙烯醇和 0.5 g OP-10 溶于 90℃ 40 mL 的蒸馏水中，时间大约 1~2h）及 10.7 mL 醋酸乙烯酯，待乳化剂全部溶解后，称 0.5 g 过硫酸铵，用 2.5 mL 水溶解于小烧杯中，将此溶液的一半倒入反应釜内，开动搅拌❶，加热恒温槽，反应温度在 65~70℃。然后用

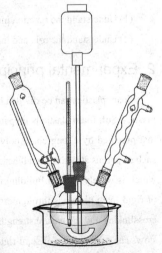

图 2.6　反应装置图

---

❶ 实验过程中控制搅拌速度恒定，否则料液乳化不完全。

fast[1]) in the dropping funnel to start the polymerization. Then add the remaining ammonium persulfate solution into the four-neck round bottom bottle, continue to heat, reflux and gradually increase the temperature[2]. Finally the temperature rises to 80 ℃ to avoid producing large amounts of foam, and keep the temperature until no reflux. Stop heating and cool the reactor to 50 ℃, then add 0.12 g of sodium bicarbonate (dissolved in 2.5 mL aqueous solution) and 4 mL of dibutyl phthalate in turn. The mixture is stirred to cool, resulting in white emulsion. It can also be diluted with water and mixed with color paste to make various colors of paint as latex paint.

Questions

(1) Compare the reaction characteristics of emulsion polymerization, solution polymerization and suspension polymerization.

(2) What are the roles of emulsifiers?

(3) What problems should be paid attention to in the experimental operation?

Experiment 10 The preparation of polyurethane foam plastics

1. The purposes

(1) Understand the reaction mechanism of preparing polyurethane foam plastics.

(2) Understand the role and influence of each component.

2. Experimental principle

Foam plastics can be divided into soft, semi-rigid, rigid foam plastics or perforated, closed-cell foam plastics. Soft foam plastic is prepared by the expansion of flexible polymer, and linear polyurethane (PU) that prepared by a long chain polyether is a typical example. Polyurethane foam has many advantages, such as porous structure, small heat capacity, low thermal conductivity, sound absorption and shockproof, which is widely used in building materials, furniture, packaging and other industries. The preparation of rigid foam plastic is through cross-linkable monomers. The ratio between compressive strength after crosslinking and the tensile strength is 0.5 or above, and elongation rate is less than 10%. The recovery is slow. The performance of semi-rigid foam plastics is between the flexible and rigid foam plastics. Open-

---

[1] Control dropping speed of vinyl acetate drops in the experiment, otherwise the agglomeration will easily appear in the emulsion.

[2] After the dropping of vinyl acetate, improve the temperature gradually.

滴液漏斗滴加 16 mL 醋酸乙烯酯（滴加速度不宜过快❶），加完后把余下的过硫酸铵加入反应瓶中，继续加热，使之回流，逐步升温❷。以不产生大量泡沫为准，最后升至 80℃，无回流为止。停止加热，冷却至 50℃后，加入 0.12 g 碳酸氢钠（溶于 2.5mL 水溶液中）再加入 4 mL 邻苯二甲酸二丁酯，搅拌冷却后，即成白色乳液。也可以用水稀释并混入色浆制成各种颜色的涂料作为乳胶漆。

思考题

（1）比较乳胶液聚合、溶液聚合、悬浮聚合的反应特点。
（2）乳化剂的作用是什么？
（3）本实验操作应注意哪些问题？

实验十　聚氨酯泡沫塑料的制备

一、实验目的

（1）了解制备聚氨酯泡沫塑料的反应原理。
（2）了解各组分的作用及影响。

二、实验原理

泡沫塑料可分为软质、半硬质、硬质的泡沫塑料或开孔、闭孔的泡沫塑料。软质的泡沫塑料是由柔韧的高聚物膨胀产生的，用长链聚醚制备的线型聚氨基甲酸酯（简称聚氨酯）就是典型的例子。聚氨酯泡沫塑料具有多孔结构，热容量小，导热系数低，吸音防震，在建材、家具、包装等行业具有广泛的应用。硬质泡沫塑料是由可交联的单体制备的，交联后的压缩强度对抗张强度之比为 0.5 或以上，伸长率小于 10%，复原慢。半硬质泡沫塑料的性能在韧性和刚性泡沫塑料之间。开孔泡沫塑料，如海绵，具有相互连通的小孔结构；闭孔

---

❶ 实验中要控制醋酸乙烯酯滴加速度，否则乳液中容易出现块状物。
❷ 醋酸乙烯酯滴加完毕后再开始逐渐升温。

cell foam plastics, such as sponges, have interconnected pore structure; Closed-cell foam plastics are wrapped by polymer and are made from disperse airbags. A certain gas is encased in viscous medium to produce porous structure. The gas can be released from polymer and also can be produced by foaming agents decomposition, such as thermal decomposition of ammonium bicarbonate to produce NH_3, CO_2 and H_2O. The gas is also generated by gasification of volatile solvent under the influence of the reaction heat, and the solvent is equivalent to a foaming agent. Polyurethane is made of automatic foaming foam plastics, and ethylene-based polymers can prepare foam plastics by mixing in a volatile substance or blowing gas into the polymer latex.

In the course of preparing polyurethane foam plastics, the reaction of isocyanates is critical. Organic isocyanates can react with any substance with active hydrogen.

$$R-N=C=O + H_2O \longrightarrow R-NHCOOH \longrightarrow RNH_2 + CO_2 \qquad (1)$$

$$R-N=C=O + R'-OH \longrightarrow R-NHCOOR' \qquad (2)$$

$$R-N=C=O + R'-NH_2 \longrightarrow R-NH-CO-NH-R' \qquad (3)$$

The formula for the chemical reaction is as follows. Among them, the isocyanate reacts with water to produce carbon dioxide gas(1), playing a foaming effect. The isocyanate reacts with alcohol (2) or amine (3), which is the polymerization reaction. Amine that generated according to formula (1) will further, react with isocyanate to generate polymer, thus the water has played a dual role in the preparation of polyurethane foam plastics.

The catalysts are also used in the process of preparing foam plastics. There are two kinds of most effective catalysts. One is some metal salts, such as divalent tin and zinc, which can activate isocyanates, especially aliphatic isocyanates; the other is tertiary amine, as a catalyst its activity of amine depends on basicity of amine and the binding force of nitrogen atom. Generally, we don't use primary and secondary amines, because they react with isocyanate to generate urea derivatives directly, which don't have enough activity to catalyze reaction. Aromatic amines and amides can not be used as the catalyst, because their alkalinity is not strong enough. The catalytic activity of aliphatic amine decreases with the increase of carbon chains.

There are hree kinds of methods for the preparation of polyurethane foam plastics in industry, namely prepolymer, semi-prepolymer and one-step method. In prepolymer method, firstly isocyanate reacts with polyol to produce prepolymer, then add water, catalysts and surfactants in the prepolymer, and make water react with the isocyanate groups. When system is foaming, chains begin to propagate (as well as the cross-linking reaction) at the same time, forming foam plastics. The semi prepolymer method is the reaction of a part of polyether or polyester polyol and all the isocyanates in the recipe to produce a prepolymer mixture with terminal isocyanate oligomers and a large amount of unreacted isocyanate. The mixture is then foamed with the remaining polyether or polyester polyol, water, catalyst and surfactant. This method is suitable for the preparation of rigid foam plastics. In one-step method, add all ingredients in a one-step to ensure that the chain propagation, gas formation and crosslinking reaction are realized almost simultaneously in a short time. The process is simple,

的泡沫塑料是由高聚物包裹起来、分散的气囊所构成。多孔的产生是黏性介质中包住了一定的气体,这种气体可由聚合物本身放出,也可加入发泡剂,通过它的分解产生。例如,碳酸氢铵受热分解产生 NH_3、CO_2 和 H_2O;还可由挥发性溶剂在反应热的影响下汽化生成,这种溶剂也相当于发泡剂。聚氨酯就是自动发泡制成的泡沫塑料;而乙烯类聚合物,可通过在其中混入挥发性物质,或在聚合物胶乳中吹入气体制得泡沫塑料。

在制备聚氨酯泡沫塑料的过程中,异氰酸酯的反应极为关键。有机异氰酸酯可与任何带有活泼氢的物质发生反应。反应式如下:

$$R-N=C=O + H_2O \longrightarrow R-NHCOOH \longrightarrow RNH_2 + CO_2 \quad (1)$$

$$R-N=C=O + R'-OH \longrightarrow R-NHCOOR' \quad (2)$$

$$R-N=C=O + R'-NH_2 \longrightarrow R-NH-CO-NH-R' \quad (3)$$

其中,异氰酸酯与水反应产生二氧化碳气体,(1)起到发泡作用异氰酸酯分别与醇(2)或胺(3)反应,即为聚合反应。式(1)中生成的胺会进一步与异氰酸酯反应生成聚合物,因此水在聚氨酯泡沫塑料制备反应中起到了双重作用。

在泡沫塑料的制备过程中也会使用催化剂,最有效的催化剂有两种:一是某些金属盐,如正二价的锡和锌,它们能够活化异氰酸酯,特别是脂肪族异氰酸酯;二是三级胺,作为催化剂,胺的活性取决于胺的碱性和氮原子的结合力。一般不用一级胺和二级胺,因为它们和异氰酸酯直接生成脲的衍生物,用脲再去催化反应是不够活泼的。芳香族的胺和酰胺也不能作为催化剂,因为它们的碱性不够强;脂肪族胺的催化活性随碳链增长而降低。

聚氨酯泡沫塑料在工业上有 3 种制备方法,即预聚体法、半预聚体法和一步法。预聚体法是先将异氰酸酯和多元醇反应生成预聚体,然后在预聚体中加入水、催化剂和表面活性剂等,使水和异氰酸酯基反应,在发泡的同时进行链增长(有的同时还有交联反应),形成泡沫塑料。半预聚体法是将一部分聚醚或聚酯多元醇和配方中全部的异氰酸酯反应生成末端带有异氰酸酯基的低聚物和大量未反应游离的异氰酸酯的预聚体混合物,该混合物再和剩余的聚醚或聚酯多元醇、水、催化剂以及表面活性剂混合进行发泡,此法适于制备硬质泡沫塑料。一步法是将所有原料一次加入,使链增长、气体生成和交联反应在短时

but the recipe must be elaborately designed. Control suitable conditions to obtain an excellent foam plastic.

In this experiment, dibutyltin dilaurate and 1,4-diazabicyclo (2,2,2) octane (DABCO) were used to catalyze the reaction of isocyanate with polyols, and soft foam plastics were prepared by one-step process.

3. Instruments and reagents

Instruments: Beaker, burette, kraft paper.

Reagents: Tolylene-2,4-diisocyanate (TDI), trihydroxy polyether (relative molecular mass is 2000-4000), dibutyltin dilaurate, deionized water, 1,4-diazabicyclo (2,2,2) octane (DABCO) (or triethanolamine), organic silicone oil.

4. Experimental procedure

A slightly harder kraft paper is folded into about 80 mm × 80 mm × 80 mm cartons to use as a mold.

Add 0.1g DABCO and 5 drops of water in a 25 mL beaker to dissolve, then add 10 g trihydroxy polyether. The resulting solution is named as A solution.

In another 250 mL beaker, 25 g of trihydroxy polyether, 10 g of toluene-2,4-diisocyanate ❶ and 5 drops of dibutyltin dilaurate are successively added into another 250 mL beaker, and the reaction heat is released at this time. This is solution B.

Add about 10 drops of organosilicon oil to the A solution. After mixing it, pour the solution into B solution quickly, stir it quickly with a glass rod❷, transfer it to the mold after the thickening of the reactants, place the mold for half an hour at room temperature, and then put it in the oven about 70 ℃ for 1h to get a piece of soft polyurethane foam plastics.

Questions

(1) Write the reaction equation of synthesizing polyurethane in this experiment.

(2) What are the nature and role of each component in the above recipe?

(3) Cut the foam plastics and observe the distribution of bubble holes. Try to discuss the various factors that affect the uniformity of bubble hole distribution.

(4) Try to analyze the reasons for uneven foaming or poorer foaming effect, and put forward to the improvement measures.

---

❶ As highly toxic drugs, tolylene-2,4-diisocyanate should be paid attention to protection when we use them. Weigh them in a fume hood and pay attention to try not to spill out. Spilled isocyanates can be dealt with 5% ammonia.

❷ The A solution should be poured into B solution quickly, and immediately stir to avoid uneven reaction.

间内几乎同时进行，工艺简单但配方必须精细设计，并控制合适的条件才能得到优良的泡沫塑料。

本实验采用二月桂酸二丁基锡和1,4-二氮杂双环（2，2，2）辛烷（DABCO）催化异氰酸酯与多元醇的反应，一步法制备软质泡沫塑料。

三、仪器和试剂

仪器：烧杯；滴管；牛皮纸。

试剂：甲苯-2,4-二异氰酸酯（TDI）；三羟基聚醚（相对分子质量2000~4000）；二月桂酸二丁基锡；去离子水；1,4-二氮杂双环（2,2,2）辛烷（DABCO）（或三乙醇胺）；有机硅油。

四、实验内容

（1）用稍硬的牛皮纸折成大约80 mm × 80 mm × 80 mm的纸盒以作模具用。

（2）在一个25 mL的烧杯中，加入0.1 g DABCO和5滴水，溶解后再加入10 g三羟基聚醚，作为A溶液。

（3）在另一250 mL的烧杯中，依次加入25 g三羟基聚醚、10 g甲苯-2,4-二异氰酸酯❶、5滴二月桂酸二丁基锡，搅匀，此时有反应热放出，此为B溶液。

（4）向A溶液中加入约10滴有机硅油，搅匀后，将此溶液迅速倒入B溶液中，用玻璃棒迅速搅拌❷，待反应物变稠后，将其转移到模具中，于室温放置30min，再放入约70℃的烘箱中烘1h，即可得到一块软质聚氨酯泡沫塑料。

思考题

（1）写出本实验合成聚氨酯的反应方程式。

（2）上述配方中各组分的性质及作用是什么？

（3）切开所制得的泡沫塑料，观察泡孔分布的情况，试讨论影响泡孔分布均匀程度的各种因素。

（4）试分析发泡不均匀或发泡效果较差的原因，并提出改进措施。

---

❶ 甲苯-2,4-二异氰酸酯为剧毒药品，使用时应注意防护，在通风橱内进行量取。注意尽量不要洒出，洒出的异氰酸酯可用5%的氨水处理。

❷ 将A溶液倒入B溶液时应迅速，并马上搅拌，以免反应不均匀。

Additional knowledge

The brief introduction of polyurethane application

Polyurethane molecules have strong polar groups, and there are hydrogen bonds in macromolecule, which give polymer high strength, wear resistance, solvent resistance and other characteristics. We can adjust the performance of polyurethane in large scale by changing monomers' structure and relative molecular mass and the like, so they have a wide range of uses in plastics (particularly foam), rubber, coatings, adhesives, synthetic fiber and other fields, and their application is still developing.

Polyurethane fiber is a kind of elastic fibers, called spandex, and they can be spinned into elastic, comfortable fiber products with other fibers.

Because of excellent adhesion of their paint film, polyurethane coating can be used to protect metal, rubber, leather, paper and wood.

Polyurethane rubbers have a particularly good wear resistance, tear strength, resistance to ozone, ultraviolet ray and oil, therefore they are used to produce tires of automobile and aircraft.

Polyurethane thermoplastic elastomers have the elasticity of rubber and easy workability of plastic, and they can be processed by machining method of thermoplastic resins, such as injection, extrusion, calendering. They also have excellent oil resistance, wear resistance, low temperature flexibility and anti-aging properties, therefore they are widely used. They can be used as auto bearing bush, bearing, tractors' crawler belt and high-speed conveyor belts for the textile industry, ect.

Because polyurethane adhesive has some active groups in molecular chains, which let it have a higher polarity, it has high adhesion properties for a variety of materials. It not only can glue porous material, such as foam plastics, ceramics, wood, fabric, etc, but also can glue the materials that have smooth and clean surface, such as steel, aluminum, stainless steel, metal foil, glass and rubber, ect. At the same time, the cementing between porous materials and the materials having smooth and clean surface is also good.

Polyurethane foam plastics can be divided into soft foam plastics and hard ones, which relates to the raw materials, synthetic process and use requirements. They have a lot of performances of heat preservation, heat insulation and sound insulation. Soft foam plastics can be used as sound insulation parts, chair cushion, clothing, precision instruments and meter packaging materials, sponge, etc; rigid foam plastics are used generally in the refrigerator, cold storage, construction, insulating material and other industrial parts.

补充知识

聚氨酯的应用

聚氨酯分子中具有较强的极性基团，在大分子中存在着氢键，使聚合物具有高强度、耐磨、耐溶剂等特点；而且可通过改变单体的结构、分子量等，在很大范围内调节聚氨酯的性能，使之在塑料（特别是泡沫塑料）、橡胶、涂料、黏合剂、合成纤维等领域中具有广泛的用途，且其用途还在不断开发中。

聚氨酯类纤维是一种具有适宜弹性的纤维，称为氨纶，可与其他纤维纺成具有弹性、舒适的纤维制品。

聚氨酯涂料由于其漆膜的黏附性很好，可用来保护金属、橡皮、皮革、纸张和木材。

聚氨酯橡胶具有特别好的耐磨性、撕裂强度、耐臭氧、紫外线和油，因此用来生产汽车和飞机轮胎。

聚氨酯热塑性弹性体，既有橡胶的弹性，又有塑料的易加工性，能用热塑性树脂的加工方法加工，如注射、挤出、压延等。它同样具有卓越的耐油性、耐磨性、低温弹性和耐老化等性能，用途广泛。可用作汽车的轴瓦、轴承、拖拉机的履带和纺织工业中的高速传送带等。

聚氨酯黏合剂，由于其分子链具有一些活泼基团和较高的极性，因而对多种材料具有极高的黏附性能。不仅可以胶接多孔材料，如泡沫塑料、陶瓷、木材、织物等，而且可以胶接表面光洁的材料，如钢、铝、不锈钢、金属箔、玻璃以及橡胶等。同时对多孔材料与表面光洁材料相互之间的胶接也是很好的。

聚氨酯泡沫塑料有软质和硬质之分，与所用原料、合成工艺以及用途要求有关。它们具有保温、绝热和隔音等性能。软质泡沫塑料用作隔音制件、椅垫、衣服、精密仪器和仪表的包装材料、海绵等；硬质泡沫塑料多用在冰箱、冷藏、建筑、绝缘材料等工业制件中。

Experiment 11 Cationic polymerization of styrene

1. The purposes

(1) Deepen understanding for the basic principle of cationic polymerization.

(2) Master experimental methods of styrene cationic polymerization triggered by boron trifluoride ether solution.

2. Experimental principle

Most of vinyl monomers can participate in radical polymerization, but they have a higher selectivity for ionic polymerization. The monomers with electron-donating groups are usually in favor of cationic polymerization, and monomers with electron-withdrawing groups are suitable for anionic polymerization. While monomers having a conjugated system can be fit for both cationic polymerization and anionic polymerization, styrene is a typical example. The mechanism of cationic polymerization of styrene is shown in Figure 2.7.

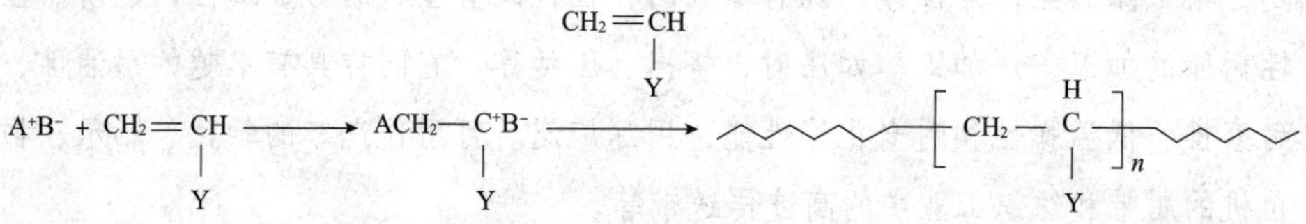

Figure 2.7 The mechanism of cationic polymerization of styrene.

As the name suggests, the propagating chains' active centers of cationic polymerization are cations, which can be carbocations, oxygen positive ions and quaternary ammonium ions. In addition to the above-mentioned monomers containing electron-donating group and a conjugated system, unsaturated compounds containing oxygen, nitrogen and ring compounds can also carry through cationic polymerization.

Initiators (also called catalysts) of cationic polymerization are electrophilic reagents. We can initiate it by cationic initiators and charge transfer complexes. Lewis acid is the most common initiator. Same as anionic polymerization, the initiation and chain propagation of cationic polymerization is fast. Cationic polymerization is more prone to chain transfer reaction in various forms, which also is the main method of chain termination. The real kinetic chain termination is less, but because of the poor stability of carbocations, Cationic polymerization is not so easy as anionic polymerization to form a living polymerization. It utilizes the interaction between specific antiparticles and carbocations to improve the stability of polymerization, thus achieve the living

实验十一　苯乙烯的阳离子聚合

一、实验目的

（1）加深理解阳离子型聚合反应的基本原理。

（2）掌握用三氟化硼乙醚溶液引发苯乙烯的阳离子型聚合反应的实验方法。

二、实验原理

多数烯类单体都能进行自由基聚合，但对于离子型聚合却有较高的选择性，通常带有推电子基团的单体有利于阳离子聚合；带有吸电子基团的单体有利于进行阴离子聚合；具有共轭体系的单体既能阳离子聚合，也能阴离子聚合，苯乙烯就是一个典型的例子。苯乙烯阳离子聚合示意图如图 2.7 所示。

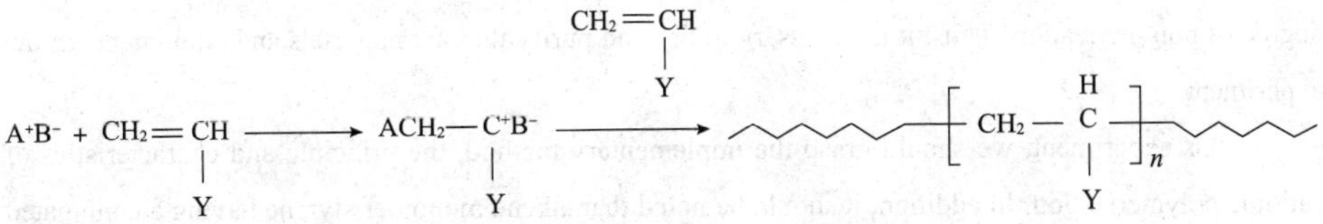

$$A^+B^- + CH_2=CH\underset{Y}{|} \longrightarrow ACH_2-C^+B^-\underset{Y}{|} \longrightarrow \sim\!\!\sim\!\!CH_2\!\!\sim\!\!\sim \left[CH_2-\underset{Y}{\overset{H}{\underset{|}{C}}}\right]_n \sim\!\!\sim$$

图 2.7　苯乙烯阳离子聚合示意图

顾名思义，阳离子聚合的增长链活性中心是阳离子，它可以是碳阳离子、氧正离子，也可以是季铵离子。可以进行阳离子聚合的单体除了上述含推电子基的单体和共轭单体外，还有含氧、氮杂原子的不饱和化合物及环状化合物进行开环聚合。

阳离子聚合的引发剂（也称催化剂）都是亲电试剂。引发方式有两种，一种是引发剂阳离子引发，另一种是电荷转移络合物引发，最常用的引发剂是 Lewis 酸。与阴离子聚合相同，阳离子聚合引发速度很快，而且链增长也很快，阳离子聚合比较容易发生链转移，反应形式多样，也是最主要的链转移终止方式。虽然真正的动力学链终止比较少，但是碳阳离子的稳定性差，并不像阴离子聚合那样容易形成活性聚合，只是利用特定的反离子与碳阳离子之间的相互作用，才可以提高其稳定性，实现活性聚合。目前活性阳离子聚合是高分

polymerization. At present, living cationic polymerization is frontier topic in the field of polymer chemistry.

When using Lewis acids as initiators, in addition to vinyl monomers, we must add cocatalyst (such as water, alcohol, certain acids and alkyl halide, etc.) for other alkene monomers, so that cationic polymerization can occur (Figure 2.8). When water and alcohol are used as cocatalyst, they form complexes with Lewis acids, which rearrange to produce cations. When alkyl halide serves as cocatalyst, initiator of the reagents is alkyl cation (carbocation).

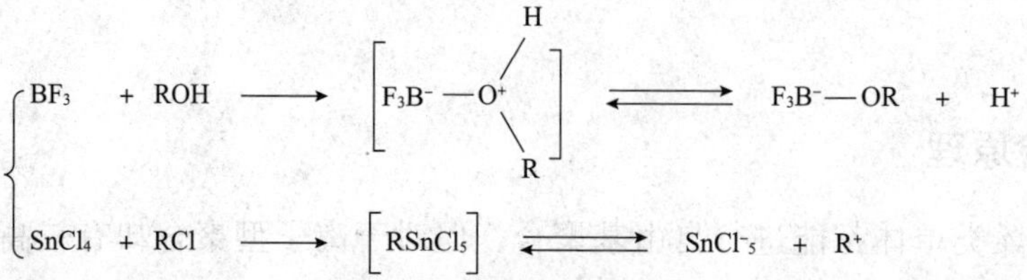

$$\begin{cases} BF_3 + ROH \longrightarrow \left[F_3B^- - O^+ \overset{H}{\underset{R}{}} \right] \rightleftharpoons F_3B^- - OR + H^+ \\ SnCl_4 + RCl \longrightarrow [RSnCl_5] \rightleftharpoons SnCl_5^- + R^+ \end{cases}$$

Figure 2.8 Interaction mechanism of initiator and cocatalyst

Cationic polymerization is sensitive to impurities, which will accelerate the reaction or inhibit polymerization. In addition, impurities can also cause chain transfer or chain termination, which decrease degree of polymerization. Thus, it is necessary to dry and purify the raw materials and equipments in the experiment.

In this experiment, we should grasp the implementary method, the principle and characteristics of cationic polymerization. In addition, it should be noted that alkene monomer styrene having a conjugated structure, can synthesize polystyrene and their copolymerization products by free radical polymerization (comprising various implementary methods), anionic polymerization and cationic polymerization, and different polymerization methods have their own characteristics and application, which is one of the reasons for wide applications of polystyrene and copolymer products and the earliest to realize industrialization.

3. Instruments and reagents

Instruments[❶]: Single-necked flask, syringes (with needles), the flanging rubber-stopper, beakers, Buchner's funnel and suction flask.

Reagents: Toluene and styrene that dried with calcium hydride and dealed with vacuum distillation, methanol, boron trifluoride ether solution ($BF_3 \cdot C_2H_5OC_2H_5$).

---

❶ The experimental materials must be dried well in advance, and laboratory glasswares are pumped and blowed repeatedly, removing the trace water and oxygen.

子化学领域中的前沿课题。阳离子聚合的特点是快引发、快增长、易转移、难终止。

采用 Lewis 酸作为引发剂时，除乙烯类单体外，其他烯类单体必须添加助催化剂（如水、醇、某些酸和卤代烷等），才能发生阳离子聚合如图 2.8 所示。当用水和醇作为助催化剂时，它们与 Lewis 酸形成络合物，再重排产生阳离子；当卤代烷作为助催化剂时，试剂的引发剂是烷基阳离子（碳阳离子）。

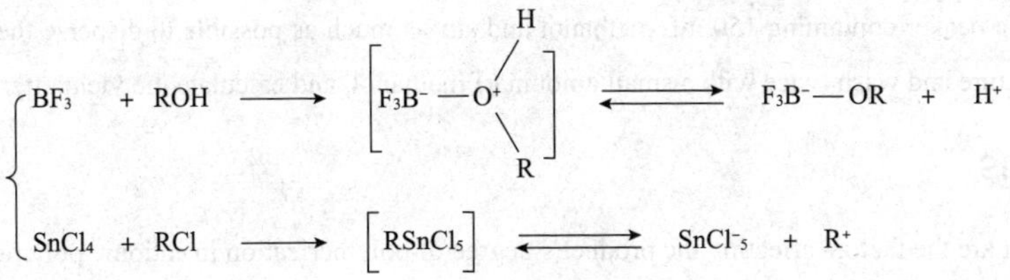

$$BF_3 + ROH \longrightarrow \left[F_3B - \overset{H}{\underset{R}{O^+}} \right] \rightleftharpoons F_3B-OR^- + H^+$$

$$SnCl_4 + RCl \longrightarrow [RSnCl_5] \rightleftharpoons SnCl_5^- + R^+$$

图 2.8 引发剂和助催化剂共同作用机理

阳离子聚合对杂质非常敏感，杂质可能会加速反应，也可能对反应起阻聚作用。此外，杂质还能引起链转移或终止，使聚合度降低。因此实验中必须仔细对原料和仪器进行干燥与除杂。

本实验的苯乙烯阳离子聚合，要求掌握阳离子聚合的实施方法以及阳离子聚合的原理和特点。此外应该注意的是，苯乙烯这一具有共轭结构的烯类单体，可以通过自由基聚合（包含多种实施方法）、阴离子聚合和阳离子聚合得到聚苯乙烯及其共聚产物，而且不同的聚合方法具有各自的特点和用途，这也正是聚苯乙烯及其共聚物产品用途多样化以及最早实现工业化的原因之一。

三、仪器与试剂

仪器❶：单口烧瓶；注射器（配针头）；翻口橡皮塞；烧杯；布氏漏斗及抽滤瓶。

试剂：经氢化钙干燥并减压蒸馏的甲苯和苯乙烯；甲醇；三氟化硼乙醚溶液（$BF_3 \cdot O(C_2H_5)_2$）。

---

❶ 实验前所用原料必须提前干燥好，实验用玻璃仪器要仔细经多次抽排，除去微量水和氧气。

4. Experimental procedure

Take a 100 mL flask which is washed and dried, vacuum it with double pipe line system❶, pass the dry high-purity nitrogen, and then vacuum it three times, and plug the bottle mouth with a flap rubber stopper. Add 60 mL of dry toluene and 15 mL of anhydrous styrene successively by syringe, and cool the system to about 20 ℃ by the cold-water bath. Then add 0.1 mL of $BF_3 \cdot C_2H_5OC_2H_5$ with a syringe, and shake flask gently to mix the reactants. Pay attention to the temperature change of the reactants and keep the temperature at 35 ~ 40 ℃ for 1.5 h after the reaction is stable. After the reaction, pour the polymer solution into a beaker containing 150 mL methanol and stir as much as possible to disperse the polymer. Filter the mixture and wash twice with a small amount of methanol, and calculate the yield after drying.

Questions

(1) What are the factors affecting the product's degree of polymerization in cationic polymerization?

(2) Compare the similarities and differences between the anionic polymerization and cationic polymerization of styrene.

Additional knowledge

The practical application of cationic polymerization

Preparing polyisobutylene and butyl rubber by isobutylene polymerization are the most important industrial applications of cationic polymerization. Polyisobutylene of low molecular weight ($\overline{M_n}$<50,000) is a viscous liquid or very sticky semi-solid, which is used as adhesives, caulking materials, sealing materials, additives of bottled fuel to improve the viscosity. Polyisobutylene of high molecular weight ($\overline{M_n} = 5 \times 10^{4-6}$) is a rubber solids, which is used in making wax and additives of other polymers and packaging materials. Relative molecular weight is controlled mainly by temperature. If $AlCl_3$ is used as initiator, the polymerization at 0 ~ -40℃ can obtain low molecular weight products, however, the polymerization at a lower temperature (-100 ℃) can obtain high molecular weight products.

Butyl rubber is the copolymer of isobutylene and small amounts of isoprene (1% - 6%). Using $AlCl_3$ as initiator and methyl chloride as diluent, methyl chloride solutions of monomers and initiators are cooled to the reaction temperature, which are injected into the reactor in the form of a thin stream, and cationic polymerization occurs at -100 ℃, which is almost instantaneous. Polymers are precipitated from the methyl chloride in the form of fine powder. The slurry is sent into hot water flash tank to remove the methyl chloride and unreacted monomers. Then add a small amount of zinc stearate (anti-condensation), sodium hydroxide, and antioxidants. The slurry of butyl rubber is filtered and dehydrated, and then it is

---

❶ When cleaning with a double pipe system, we should be careful not to let the nitrogen flow too big. In order to eliminate the trace moisture, we also can use a hair dryer or gas burner to bake the bottom of bottle when vacuuming.

四、实验内容

取一洗净烘干的 100 mL 烧瓶,用双排管系统抽真空❶,通干燥的高纯氮气,如此反复进行抽排三次后抽真空,用翻口橡皮塞塞住瓶口。用注射器依次加入 60 mL 无水甲苯、15 mL 无水苯乙烯,用冷水浴将体系降温至 20℃左右时,用注射器加 0.1 mL $BF_3 \cdot O(C_2H_5)_2$,轻轻摇动烧瓶使反应物混合均匀,注意反应物温度的变化,控制其温度保持在 35~40℃,待反应平稳后,放置 1.5 h。反应结束后,将聚合物溶液倒入装有 150 mL 甲醇的烧杯中,同时要尽量搅拌,以使聚合物分散。过滤,用少量甲醇洗涤两次,烘干后计算产率。

思考题

(1) 在阳离子聚合反应中影响产物聚合度的因素有哪些?
(2) 对比苯乙烯的阴离子聚合和阳离子聚合的异同。

补充知识

阳离子聚合的应用

异丁烯聚合制备聚异丁烯和丁基橡胶是阳离子聚合最主要的工业应用。低相对分子质量($\overline{M_n}$<50000)的聚异丁烯是黏滞液体和很黏的半固体,用作黏结剂、嵌缝材料、密封材料、桶装油料的添加剂以改进黏度。分子量高的聚异丁烯($\overline{M_n}$=5×10^{4~6})是橡胶状固体,用于制蜡以及其他聚合物和封装材料的添加剂。分子量主要靠温度来控制,若以 $AlCl_3$ 作引发剂,在 0~-40℃聚合,得到分子量低的产物;在更低温(-100℃)聚合时,得到相对分子质量高的产物。

丁基橡胶是异丁烯和少量异戊二烯(1%~6%)的共聚物。以 $AlCl_3$ 作引发剂,以氯甲烷为稀释剂,单体和引发剂的氯甲烷溶液分别冷却到反应温度,以细流注入反应器,在 -100℃进行阳离子聚合,聚合几乎瞬时完成。聚合物从

---

❶ 在使用双排管系统进行除杂时,要小心操作,氮气通入时流量不能太大。在抽真空时还可以用电吹风或煤气喷灯烘烤瓶底,尽量排除微量水分。

sent to the extruder for drying and extrusion. Finally, it can be packed ino butyl rubber products.

The content of isoprene in butyl rubber depends on the degree of crosslinking required, and the relative molecular weight is at least 200000. Butyl rubber does not crystallize when cooled, and remains soft at - 50 ℃ . It has the advantages of weather resistance, ozone resistance and good air tightness. 75% of its output is used as inner tube.

Isoprene content of butyl rubber depends on the required crosslinking degree. Butyl rubber, whose relative molecular weight is more than 200,000 kDa, is not sticky. Butyl rubber will not crystallize when it is cooled and even remains soft at -50 ℃, which has many advantages, such as, weather resistance, ozone resistance and good air tightness. 75% of its output is used as inner tube of tire.

Experiment 12 Reversible addition-fragmentation chain transfer (RAFT) polymerization of styrene

1. The purposes

(1) Understand the nature and types of living free radical polymerization.
(2) Understand the mechanism of RAFT polymerization.
(3) Learn the operation of reaction without oxygen and water in Schlenk tube.

2. Experimental principle

(1) Experimental principle and background

Free radical polymerization is a powerful tool to prepare polymers in industry. 70% of vinyl polymers and 50% of plastics are synthesized by free radical polymerization in the world. In conventional free radical polymerization, slow initiation, fast propagation, easy chain termination and chain transfer lead to short lifetime of free radicals (about 1 s). So it is hard to handle the molecular structure and molecular weight of product by the conventional free radical polymerization. Polydispersity of molecular architecture greatly affects the performance of polymers. In addition, it is impossible to obtain polymers with defined molecular architecture via the conventional free radical processes. Therefore, it is significant to develop living free radical polymerization.

In the past decades, living radical polymerization (LRP) has achieved great development. Several successful LRP systems have been discovered, including Iniferter Radical Polymerization, Stable Free Radical Polymerization system (SFRP), Atom Transfer Radical Polymerization (ATRP) and Reversible addition-fragmentation chain transfer (RAFT) polymerization. Among them, RAFT polymerization achieves the control for free radical concentration by reversible chain transfer of propagating free

氯甲烷中以细粉状沉淀出来。将淤浆液送入热水闪蒸槽，脱除氯甲烷和未反应的单体。再加入少量硬脂酸锌（防凝聚）、氢氧化钠和抗氧剂。将丁基胶的淤浆过滤脱水，送入挤出机干燥并挤出，即可包装成丁基胶商品。

丁基胶中异戊二烯的含量视所需要的交联度而定，分子量至少在20万以上才不发黏。丁基胶冷却时也不结晶，-50℃也保持柔软，具有耐候性、耐臭氧和气密性好等优点，其产量的75%用作轮胎内胎。

实验十二　苯乙烯的可逆加成-断裂链转移（RAFT）聚合

一、实验目的

（1）了解活性自由基聚合的本质和类型。
（2）理解RAFT聚合的机理。
（3）掌握用聚合管进行无水无氧反应的操作。

二、实验

（1）实验背景

自由基聚合在工业生产中是一种应用最广泛和最重要的合成聚合物的手段，世界上70%以上的烯类聚合物以及50%以上的塑料源于自由基聚合。但是在传统自由基聚合中，会出现引发慢，增长快，易发生链终止和链转移等现象，链自由基的寿命很短（1 s左右），导致产物的分子结构和分子量难以控制，结构的多分散性较高，从而严重影响了聚合物的性能。此外，传统的自由基聚合也不能用于合成指定结构的规整聚合物。因此，实现自由基聚合的活性化具有十分重要的意义。

近年来，活性自由基聚合得到了很大的发展，形成了几种比较成功的可控/活性自由基聚合体系，主要有Iniferter自由基聚合体系、稳定自由基聚合体系、原子转移自由基聚合体系（Atom Transfer Radical Polymerization，ATRP）和可逆加成-断裂链转移聚合体系（Reversible Addition and Fragmentation Chain

radicals and thio ester compound, thus achieveing the control for the polymerization reaction. Almost all vinyl monomers can be polymerized by RAFT. Besides polymerization and bulk polymerization, polymerization can also be carried out by suspension polymerization and emulsion polymerization using water as a reaction medium.

As shown in Figure 2.9, the RAFT polymerization mechanism begins to forming a radical (I·) that derives from an initiator, which reacts with monomer (M) to produce a propagating radical (P_n·). P_n· reacts reversibly with chain transfer agent, which produces the stable radical intermediate, but P_n· could not react with monomer. Fragmentation of the intermediate radical provides a temporarily deactivated dormant polymer [P_n-S-C(=S)-Z] and a new radical (R·) that re-initiates the reaction. This radical (R·) reacts with a monomer, forming a new propagating radical (P_m·). The end of P_n-S-C (= S) –Z contains di-thiocarbonyl structure, which can also be used as a chain transfer agent to react with other living radical P_m·, then forming free radical intermediates, which further decompose to generate free radical P_n and P_m-S-C (= S) -Z. Thus, a new rapid reversible equilibrium is established, allowing us to control the concentration of propagating free radical at a low level in polymerization system. After a sufficient long time, molecular weights of P_m and P_n tend to equality, thus realizing control for the activity growth process, so we can synthesize the polymers with narrow molecular weight distribution.

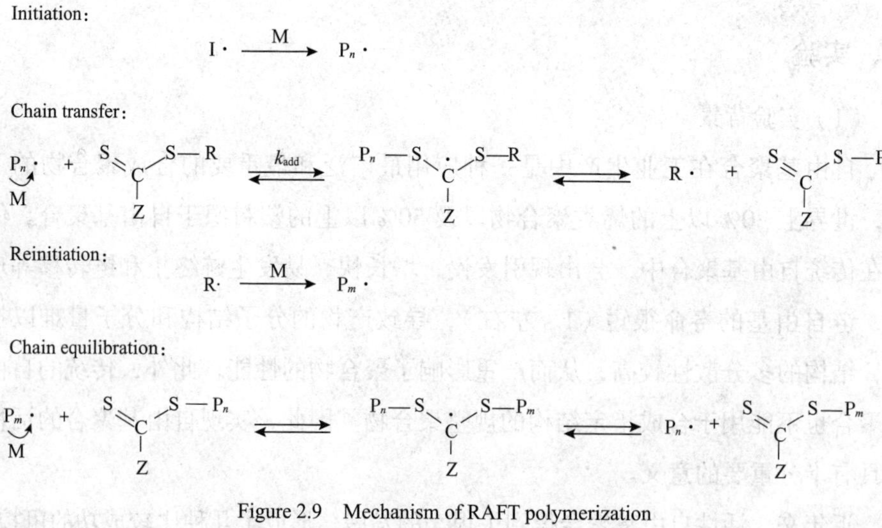

Initiation:

$$I \cdot \xrightarrow{M} P_n \cdot$$

Chain transfer:

Reinitiation:

$$R \cdot \xrightarrow{M} P_m \cdot$$

Chain equilibration:

Figure 2.9　Mechanism of RAFT polymerization

Chain transfer agent (CTA) is the key to realize the RAFT living polymerization, so it is important to choose a suitable CTA in RAFT polymerization. The CTA with high chain transfer constant and specific structure is preferred. There are some typical CTAs shown in Figure 2.10.

Transfer，RAFT）其中，RAFT 聚合是通过增长自由基与硫代酯类化合物的可逆链转移来实现对自由基浓度的控制，从而实现对聚合反应的控制。几乎所有的烯类单体都可以进行 RAFT 聚合，聚合方法除可采用溶液聚合和本体聚合外，还可用水作为反应介质进行悬浮和乳液聚合。

RAFT 聚合的机理如图 2.9 所示。引发剂裂解产生的自由基 I· 和引发单体聚合生成链增长自由基 P_n·，它们能够与链转移剂发生可逆的链转移反应，形成稳定的自由基中间体，但不能和单体发生反应。该自由基中间体可以裂解产生可再次与单体引发反应的新的自由基 R· 和暂时失活的休眠聚合物 P_n—S—C（=S）—Z。R· 可以引发单体聚合形成链自由基 P_m·，而聚合物 P_n—S—C（=S）—Z 的末端含有二硫代羰基结构，它同样可以作为链转移剂与其他活性自由基 P_m· 发生反应，再形成自由基中间体，然后进一步分解生成自由基 P_n· 和聚合物 P_m—S—C（=S）—Z。这样，一个新的快速可逆的平衡就建立起来了，从而可以控制聚合体系中的增长自由基的浓度维持在一个较低的水平。经过足够的时间进行反应及平衡后，P_m· 与 P_n· 的分子量趋近于相等，从而实现了对活性增长过程的控制，因此可以得到分子量分布较窄的聚合物。

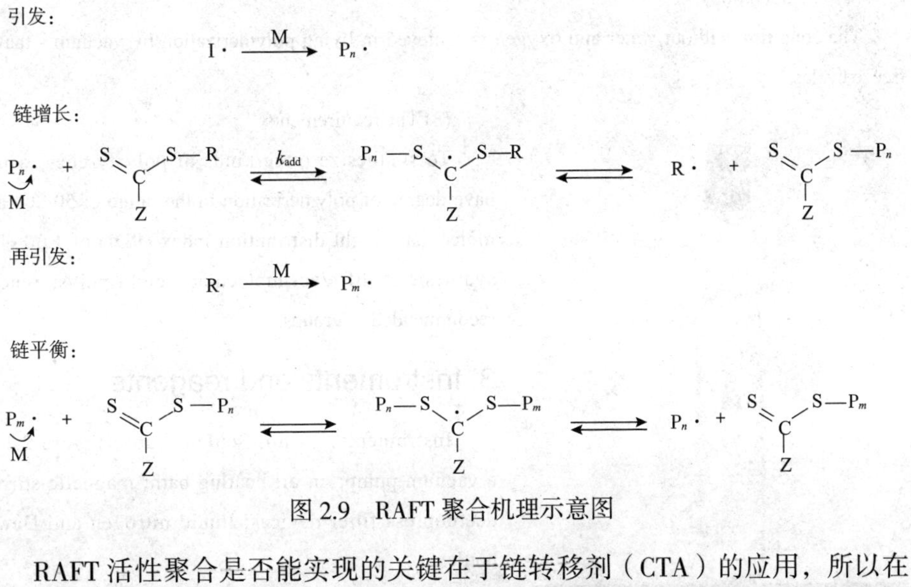

引发：

$$I· \xrightarrow{M} P_n·$$

链增长：

$$P_n· \underset{M}{} + \underset{Z}{S=C-S-R} \xrightleftharpoons{k_{add}} P_n-S-\underset{Z}{C·}-S-R \rightleftharpoons R· + \underset{Z}{S=C-S-P_n}$$

再引发：

$$R· \xrightarrow{M} P_m·$$

链平衡：

$$P_m· \underset{M}{} + \underset{Z}{S=C-S-P_n} \rightleftharpoons P_n-S-\underset{Z}{C·}-S-P_m \rightleftharpoons P_n· + \underset{Z}{S=C-S-P_m}$$

图 2.9　RAFT 聚合机理示意图

RAFT 活性聚合是否能实现的关键在于链转移剂（CTA）的应用，所以在反应体系中链转移剂的选择十分关键。具有高链转移常数和特定结构是链转移剂的必要因素，常用的链转移剂如图 2.10 所示。

Dithioester

1a Z=Ph, R=C(CH$_3$)$_2$Ph
1b Z=Ph, R=CH(CH$_3$)Ph
1c Z=Ph, R=CH$_2$Ph
1d Z=Ph, R=C(CH$_3$)(CN)CH$_2$CO$_2$Na

1e Z=Ph, R=C(CH$_3$)$_2$CN
1f Z=CH$_3$, R=CH$_2$Ph
1g Z=Ph, R=C(CH$_3$)(CN)CH$_2$CH$_2$OH
1h Z=Ph, R=C(CH$_3$)(CN)CH$_2$CH$_2$CO$_2$H

Trithioester

Figure 2.10 Typical chain transfer agents

(2) The operation

The condition without water and oxygen is achieved in living polymerization by vacuum - thaw - freeze cycles.

Figure 2.11 Schlenk tube

(3) The requirements

To synthesize three kinds of polystyrenes, which have degree of polymerization in the range of 50-200 and molecular weight distribution index (PDI) of 1.30. The synthesis of different molecular weight polystyrene is recommended in groups.

3. Instruments and reagents

Instruments: 10 mL Schlenk tube (Figure 2.11); a vacuum pump; an oil heating bath; magnetic stirrer; decompress filter devices; liquid nitrogen and Dewar bottle.

Reagents: S-dodecyl-S'- (α, α- dimethyl acetoxy) trithiocarbonate chain transfer agent (DDMAT, as shown

二硫代酯

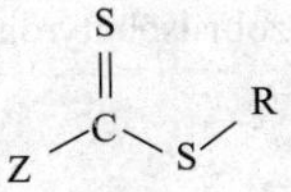

1a Z=Ph, R=C(CH$_3$)$_2$Ph
1b Z=Ph, R=CH(CH$_3$)Ph
1c Z=Ph, R=CH$_2$Ph
1d Z=Ph, R=C(CH$_3$)(CN)CH$_2$CO$_2$Na

1e Z=Ph, R=C(CH$_3$)$_2$CN
1f Z=CH$_3$, R=CH$_2$Ph
1g Z=Ph, R=C(CH$_3$)(CN)CH$_2$CH$_2$OH
1h Z=Ph, R=C(CH$_3$)(CN)CH$_2$CH$_2$CO$_2$H

三硫代酯

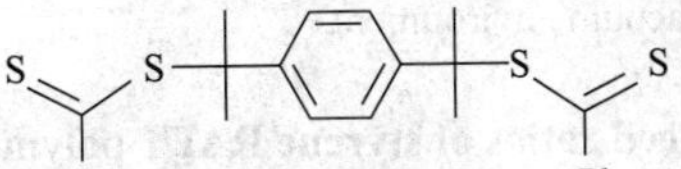

图 2.10 常用的链转移剂

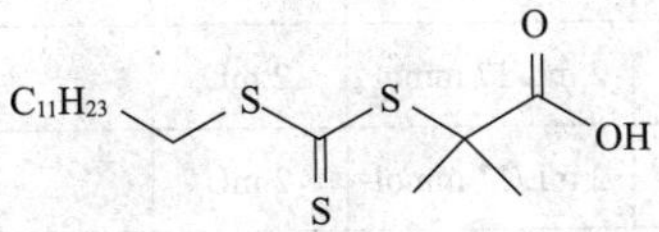

（2）实验条件

活性聚合中无水无氧条件的实现：抽真空 – 解冻 – 冷冻循环。

（3）实验要求

合成出聚合度在 50~200 范围内的三种聚苯乙烯，分子量分布指数 1.30，建议分组进行不同分子量聚苯乙烯的合成实验。

三、仪器与试剂

仪器：10 mL Schlenk 聚合管，如图 2.11 所示；真空泵；油浴锅；搅拌器；抽滤装置；液氮和杜瓦瓶。

试剂：S- 十二烷基 -S' -（α, α- 二甲基乙酸基）三硫代碳酸酯链转移剂（DDMAT），如图 2.11 所示；精制苯乙烯（St）；偶氮二

图 2.11 Schlenk 聚合管

in 3 of Figure 2.11); purified styrene (St); 2,2-azobisisobutyronitrile (AIBN); tetrahydrofuran (THF), absolute ethanol.

4. Experimental procedure

Add CTA, St, AIBN and THF (DDMAT/AIBN=10:1 mol/mol, THF/St=1:1 v/v) into a 10 mL polymerization tube containing a magneton (Table 2.5). The tube is sealed and connected to vacuum pump by a 2-way valve. Immerse the tube in liquid nitrogen to freeze solution. Then switch the valve to pump the frozen solution. 5 minutes later, close the valve and move the tube from liquid nitrogen to thaw the reaction solution. This freeze- vacuum-thaw cycle is repeated 3 times to degas the solution. Finally, close polymerization tube cock in vacuum environment.

Table 2.5 Feed ratios of styrene RAFT polymerization

Sample	CTA 3	AIBN	St	THF	Theoretical degree of polymerization (if conversion is 100%)
1	63 mg	2.8 mg	2 mL/17 mmol	2 mL	100
2	32 mg	1.4 mg	2 mL/17 mmol	2 mL	200
3	21 mg	0.9 mg	2 mL/17 mmol	2 mL	300

Place the sealed tube at 80 °C oil bath for 1-2 h under magnetic stirring. Cool it rapidly in ice bath to stop the polymerization. After diluting to a suitable concentration, the polymerization mixture is precipitated with a large excess of absolute ethanol, and separate the product by decompress filtration. Collecting the powder after being dried in vacuum dryer. Weigh the product and calculate the yield.

5. Questions

(1) What is the role of CTA in RAFT polymerization?

(2) How to understand the basic idea to achieve "living" radical polymerization(LRP)? What are the differences between LRP and living anionic polymerization?

(3) Please study the recent developments of RAFT polymerization via some new publications.

(4) Please list some typical RAFT systems, including CTA, initiator and monomer.

(5) Measure the relative molecular weight and molecular weight distribution of product by GPC, and calculate the average molecular weight by ^{1}H NMR.

6. Synthesis of S-dodecyl -S'- (α, α- dimethyl acetoxy) trithiocarbonate

Add 1-Dodecanethiol (20.19 g, 0.1 mol), acetone (48.14 g, 0.83 mol), and phase transfer

异丁腈（AIBN）；四氢呋喃（THF）；无水乙醇。

四、实验内容

将一定量的 CTA，St，AIBN 和 THF 按照表 2.5 中的比例（DDMAT/AIBN（摩尔比）=10∶1，THF/St（体积比）=1∶1）加入到 10 mL 的聚合管中，放入一颗磁子，将聚合管连接到真空泵上。将聚合管放入液氮中冷冻至反应液凝固，打开活塞抽真空，5min 后取出解冻，待反应液完全解冻后再放入液氮中冷冻，重复抽真空–解冻–冷冻操作 3 次，最后在抽真空环境中关闭聚合管活塞。

表 2.5 苯乙烯的 RAFT 聚合投料比例

序号	CTA/mg	AIBN/mg	St/（mL/mmol）	THF/mL	理论聚合度（转化率为100%）
1	63	2.8	2/17	2	100
2	32	1.4	2/17	2	200
3	21	0.9	2/17	2	300

将封闭的聚合管放置于 80 ℃ 的油浴中，磁力搅拌，聚合 1~2h，冰水冷却终止反应，将反应液稀释至合适浓度后逐滴加入到搅拌的乙醇溶液中，聚合物沉淀析出，抽滤收集固体，真空干燥，得到粉末状固体产物，计算产率。

五、思考题

（1）RAFT 试剂在聚合中起到哪些作用？

（2）如何理解实现"活性"自由基聚合的基本思路？它与阴离子活性聚合有什么差异？

（3）查阅资料，了解 RAFT 聚合的研究进展。

（4）列举一些其他常用的 RAFT 试剂、引发体系和适用的单体。

（5）利用凝胶渗透色谱测定产物的分子量及其分布，并利用核磁共振氢谱测定计算产物的分子量。

六、链转移剂 S– 十二烷基 –S'– (α, α'– 二甲基 –α''– 乙酸) 三硫代碳酸酯的合成

链转移剂 S– 十二烷基 –S'– (α, α'– 二甲基 –α''– 乙酸) 三硫代碳酸酯（DDMAT）的合成具体步骤如下：在氮气保护下，向 500 mL 三口圆底烧瓶中

catalyst Aliquot 336 (tricaprylylmethylammonium chloride, 1.62 g, 0.004 mol) orderly in a 500 mL three-necked flask under the nitrogen atmosphere , and cool it to 10 °C. Add sodium hydroxide solution (50%) (8.40 g, 0.105 mol) within 10 min. After stirring reaction for an additional 15 min, add carbon disulfide (7.61 g, 0.1 mol) in acetone (10.10 g, 0.17mol) within 10 min, then the color of the system turns red. Ten minutes later, add chloroform (17.81 g, 0.6 mol) quickly, then add 50% sodium hydroxide solution (40 g, 0.5 mol) dropwise within 15 min. Stir the reaction overnight, add 150 mL water, then add 25 mL concentrated HCl to acidify the aqueous solution. Remove acetone from the reaction flask under reduced pressure. Collect the solid with a Buchner funnel, then the solid is dissolved in 250 mL of 2-propanol. The insoluble product was filtered and removed. The yellow solid was obtained after the isopropanol was removed by decompression. The crude product was recrystallized with n-hexane and purified twice to obtain bright yellow product.

Experiment 13 Polymer Characterization—Viscosimetry

1. The purposes

(1) Know about the characterization of polymers.

(2) Master the methods of measuring the viscosity of polymer by Ubbelohde viscometer.

2. Experimental principle

The viscosity of polymer in dilute solution reflects the existence of internal friction in fluid flow. The higher the molecular weight of polymer is, the larger the contact surface between polymer and solvent, and therefore the higher intrinsic viscosity of polymer. The empirical relationship between intrinsic viscosity and molecular weight of polymer can be expressed by Mark-Houwink equation [formula (1)]. In the equation, M is the viscosity average molecular weight, K is the proportional constant, which is more obviously affected by temperature, while α is a parameter related to molecular properties, mainly determined by the stretching degree of polymer coil in a certain temperature solvent, and its value is between 0.5-1. The absolute values of K and α can be obtained by osmotic pressure method and light scattering method. it should be noted that viscosity method can only determine the viscosity value η.

$$\eta = KM^{\alpha} \tag{1}$$

The physical parameters involved in the experiment of viscosity method include pure solvent

依次加入十二硫醇（20.19 g，0.1 mol）、丙酮（48.14 g，0.83 mol）和相转移催化剂 Aliquot 336（三辛酰基甲基氯化铵，1.62 g，0.004 mol），降温至10℃。在10min内向反应体系中滴加50%的氢氧化钠水溶液（8.40 g，0.105 mol）。搅拌反应15min后，将二硫化碳（7.61 g，0.1 mol）溶于丙酮（10.10 g，0.17 mol），在10 min内加入反应瓶中，这时反应体系变为红褐色。10min后，迅速向瓶中加入氯仿（17.81 g，0.6 mol），接着滴加50%的氢氧化钠水溶液（40 g，0.5 mol），15min滴完，搅拌下反应过夜。向反应体系中加入150 mL水和25 mL浓盐酸，减压除去反应瓶中的丙酮，过滤收集产生的沉淀物。向沉淀中加入250 mL异丙醇溶解，把不溶的产物过滤除去后，将异丙醇减压除去后得到黄色固体。粗产物用正己烷重结晶两次纯化得到亮黄色产物。

实验十三　聚合物表征——黏度

一、实验目的

（1）简单了解聚合物的表征手段。

（2）熟练掌握利用乌氏黏度计测量聚合物黏度的方法。

二、实验原理

高聚物在稀溶液中的黏度反映了流体在流动时存在着内摩擦。高聚物的分子量越大，它与溶剂间的接触表面也越大，表现出的特性黏度也越大。特性黏度和分子量之间的经验关系可以用马克·霍温克方程（Mark-Houwink方程）表示（式（1）），其中 M 是黏均分子量，K 为比例常数，受温度影响较为明显，而 α 是与分子性状有关的参数，主要取决于高分子线团在一定温度的溶剂中的舒展程度，其数值介于 0.5~1 之间。K 和 α 与的绝对数值可以通过渗透压法、光散射法等手段得到，利用黏度法只能测定黏度数值 η。

$$\eta = KM^{\alpha} \tag{1}$$

黏度法实验中涉及到的一些物理量有纯溶剂黏度 η_0，溶液黏度 η，相对黏

viscosity η_0, solution viscosity η, relative viscosity η_r, specific viscosity η_{sp}, intrinsic viscosity $[\eta]$, reduced viscosity η_{sp}/C. In the case of infinite dilution of polymer solution ($C \rightarrow 0$), $\eta_{sp}/C = \ln\eta_r/C = [\eta]$. Therefore, if η_{sp}/C and $\ln\eta_r/C$ are plotted respectively for C, the intercept value extrapolated t₀ $C \rightarrow 0$ is $[\eta]$. The two lines should intersect at one point, which can also check the reliability of the experiment (Figure 2.12). The relevant calculation formula is as follows:

$$\eta_r = t/t_0 \tag{2}$$

$$\eta_{sp} = (t - t_0)/t_0 \tag{3}$$

$$\eta_{sp}/C = [\eta] + k' [\eta]^2 C \tag{4}$$

$$(\ln\eta_r)/C = [\eta] + k'' [\eta]^2 C \tag{5}$$

Where, t_0 and t are the efflux time of pure solvent and polymer solution, respectively.

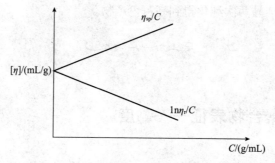

Figure 2.12 calculation of intrinsic viscosity $[\eta]$ by extrapolation

3. Instruments and reagents

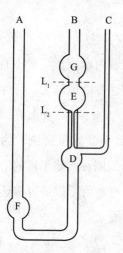

Figure 2.13 Ubbelohde viscometer

Instruments: Constant temperature bath, capable of maintaining ±0.01 ℃ at 25.0 ℃; viscometers with efflux times greater than 100 sec for the solvent (Ubbelohde viscometer, Figure 2.13); a timer with a scale of 0.1 seconds or less; volumetric flasks with stopper; pipettes; pure nitrogen at low pressure; porous glass with appropriate capacity or 1.0 μm micropore filter.

Reagents: Poly (methyl methacrylate) (PMMA) or polystyrene (PS) samples prepared in advance; chloroform or toluene.

4. Experimental procedure

Prepare chloroform solution with PMMA concentration of 0.5 mg/mL or toluene solution with PS concentration of 1.0 mg/mL in advance. Take toluene solution of polystyrene as an example,

度 η_r，增比黏度 η_{sp}，特性黏度 $[\eta]$，比浓黏度 η_{sp}/C。在高分子溶液无限稀释的情况下（$C \to 0$），$\eta_{sp}/C = \ln\eta_r/C = [\eta]$，所以将 η_{sp}/C、$\ln\eta_r/C$ 分别对 C 作图，外推到 $C \to 0$ 的截距值就是 $[\eta]$。两根线应会合于一点，这也可校核实验的可靠性（图2.12），相关的计算公式如下：

$$\eta_r = t/t_0 \tag{2}$$

$$\eta_{sp} = (t - t_0)/t_0 \tag{3}$$

$$\eta_{sp}/C = [\eta] + k'[\eta]^2 C \tag{4}$$

$$(\ln\eta_r)/C = [\eta] + k''[\eta]^2 C \tag{5}$$

式中，t 为高分子溶液流出时间，t_0 为纯溶剂流出时间。

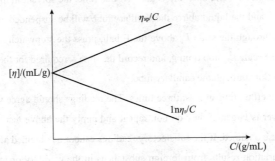

图2.12　外推法求特性黏度 $[\eta]$

三、仪器和试剂

仪器：恒温水浴，25 ℃时保温误差在 ±0.01 ℃；溶液的流出时间远大于 100s 的黏度计（乌氏黏度计，图2.13）；刻度在 0.1s 或是以下的计时器；带有塞子的容量瓶；吸量管；氮气；容量适宜的多孔玻璃或是 1.0 μm 的微孔过滤器。

药品：聚甲基丙烯酸甲酯（PMMA）或聚苯乙烯样品；氯仿或甲苯。

四、实验步骤

提前准备 PMMA 浓度为 0.5 mg/mL 的氯仿溶液或是聚苯乙烯浓度为 1.0 mg/mL 的甲苯溶液。以聚苯乙烯的甲苯溶液

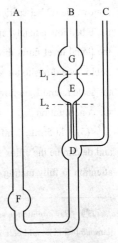

图2.13　乌氏黏度计

accurately weigh 25 mg PS into a 25 mL volumetric flask, add appropriate amount of toluene, cover the plug and gently shake; do not shake violently, because polymer particles are easy to adhere to the wall above the liquid level and cannot be dissolved. Repeat shaking until the polymer is completely dissolved, and dilute to volume with toluene. After the polymer is completely dissolved, filter the solution with a needle filter for use ❶.

(1) Rinse the viscometer with solvent and pour the waste liquid into the water tank. Fix the viscometer in a 25 ℃ constant temperature water bath and make sure that the vertical tube is vertical and G ball is immersed in water. C pipe is covered with a rubber pipe.

(2) At the end of heating (at least 10 min), accurately measure 10 mL of pure solvent with a pipette, and inject it into the viscometer through tube A, and start the determination after 10-15 min of incubation. Clamp the C tube and suck the pure solvent to 1/2 of the G ball with the rubber suction bulb in tube B, remove the rubber suction bulb and open the C tube. At this time, the solvent in the D ball will flow into the F ball immediately, and the liquid above the capillary tube will be suspended and begin to fall. When the liquid level flows through the scale L_1 above the E ball, press the stopwatch immediately. When the liquid level drops to the scale L_2, stop timing, and record the time t_0 required for the pure solvent between the scales L_1 and L_2 to flow through the capillary tube.

(3) Determine the efflux time at least three times. The readings should agree within 0.1 sec or 0.1% of their mean, whatever is larger. If they do not, repeat and apply the above test until three or even six satisfactory readings are obtained. If the satisfactory readings cannot be obtained after several operations, it is likely that the variation results from foreign substances in the capillary or inadequate temperature control. Locate and remedy before proceedings. Find out the mistakes, correct them in time, and remeasure them.

(4) Accurately measure 5 mL of pure polymer solution and inject it into the viscometer through tube A. Clamp the C tube and suck the pure solvent to 1/2 of the G ball with the rubber suction bulb in tube B, remove the rubber suction bulb and open the C tube. At this time, the solvent in the D ball will flow into the F ball immediately, and the liquid above the capillary tube will be suspended and begin to fall. When the liquid level flows through the scale L_1 above the E ball, press the stopwatch immediately. When the liquid level drops to the scale L_2, stop timing. Record the time required for the polymer solution between the scale L_1 and L_2 to flow through the capillary tube. Repeat three times to obtain the efflux time t_1 under the concentration.

(5) Add 5 mL, 5 mL, 10 mL and 10 mL solvent accurately from tube A in turn, dilute the solution and determine the efflux time t_2, t_3, t_4, t_5 of the solution at each concentration with the same method. Pay attention to fully mixing after adding solvent each time, and suck the solution to pass the E ball and G

---

❶ Pay attention to safety when using highly toxic organic solvents, such as chloroform or toluene; toluene is flammable and should be far away from fire sources. Use these solvents in the fume hood and use less. Avoid open fire or electric spark in the experiment. Safety glasses must be worn in the laboratory at all times.

为例，准确称量 25 mg PS 置入 25 mL 的容量瓶中，加入适量甲苯，盖上塞子，轻轻地摇动；不要剧烈摇晃，因为聚合物微粒容易附着在液面以上的杯壁上不能溶解。重复摇动直至聚合物完全溶解，用甲苯定容至刻度线。待聚合物完全溶解后用针头式过滤器过滤溶液待用❶。

（1）用溶剂冲洗黏度计并将废液倒入水槽中。将黏度计固定在 25 ℃ 恒温水浴中，固定好并确保直立管竖直，确保水面没过 G 球。将 C 管套上橡胶管。

（2）升温结束后（耗时最少 10 min），用移液管量取纯溶剂 10 mL，由 A 管注入黏度计中，恒温 10~15 min 开始测定。用夹子将 C 管夹紧，在 B 管用吸耳球将纯溶剂吸至 G 球 1/2 处，取下吸耳球并打开 C 管，此时 D 球内的溶剂会立即流入 F 球，毛细管以上的液体悬空并开始下落。当液面流经 E 球上刻度 L_1 时，立即按秒表计时，当液面降至刻度 L_2 时停止计时，记录刻度 L_1 和 L_2 之间的流体流经毛细管所需的时间 t_0。

（3）重复测量至少三次。读数的误差在 0.1 s 即 0.1% 以内，或是稍大一点。如果没有得到符合要求的读数，就重复测量，直至找到 3 个甚至 6 个满意的读数。按照上述操作找出 3 个读数。多次操作后仍得不到满意的读数，这很可能是外来物质在毛细管中或温度控制不充分的结果。找出错误并及时改正，重新测量。

（4）准确量取 5 mL 纯净的聚合物溶液，从管 A 注入球状玻璃管 F 中。用夹子将 C 管夹紧，在 B 管用吸耳球将聚合物溶液吸至 G 球 1/2 处，取下吸耳球并打开 C 管，此时 D 球内的聚合物溶液会立即流入 F 球，毛细管以上的液体悬空并开始下落。当液面流经 E 球上刻度 L_1 时，立即按秒表计时，当液面降至刻度 L^2 时停止计时，记录刻度 L_1 和 L_2 之间的流体流经毛细管所需的时间，重复 3 次得到该浓度下的流出时间 t_1。

（5）依次由 A 管准确加入 5 mL、5 mL、10 mL、10 mL 溶剂，稀释溶液，用同样的方法测定每种浓度下溶液的流出时间 t_2、t_3、t_4、t_5。注意每次加入溶剂后要充分混匀，并抽洗黏度计的 E 球和 G 球，保证各处黏度相同。

---

❶ 在使用高毒性的有机溶剂，如氯仿、甲苯等，要注意安全；其中甲苯是易燃物，要远离火源，在通风橱中使用且用量要少。在实验中避免出现明火或是电火花。在实验过程中要佩戴护目镜。

ball of viscometer to ensure the same viscosity.

(6) After the experiment, remove the viscometer from the water bath, empty the viscometer and rinse with pure solvent. Pour out the waste liquid and blow dry the viscometer with pure nitrogen.

Questions

(1) Include a table of values of C, η_r, η_{sp}, η_{sp}/C, and η_{inh}, and the graph prepared. Report $[\eta]$, k' k'' the solvent, and the temperature of measurement.

(2) Calculate the average molecular weight using the Mark-Houwink equation (search for K and α from literature).

Experiment 14 Condensation polymerization: preparation of nylon 66

1. The purposes

(1) Synthesize nylon 66, and calculate the yield.

(2) Grasp the fundamental concepts and theories involved with nylon synthesis.

(3) Understand the characteristics of condensation polymerization.

2. Experimental principle

Polymerization reactions may be broadly classified as step-growth and chain-growth reactions. In step-growth reactions, polymer chains grow in a slow, step-wise manner; while in chain-growth, reactions' polymer chains grow rapidly to high molecular weight. Condensation and some ring-opening polymerizations are examples of step-growth reactions. Free radical and UV initiated polymerizations are examples of chain-growth reactions.

Condensation polymerization reactions involve the loss of a small molecule, such as water. The following are some examples of reactions to afford common step-growth polymers (Figure 2.14).

For condensation polymerization to occur, each reacting molecule must have at least two reactive groups (functionality of at least 2). Functionality of greater than 2 results in branching or cross-linking. To obtain high molecular weight polymer, it is generally necessary to initiate the reaction with a stoichiometric ratio of reacting groups. The reaction mixture should be free of impurities with similar functionality, and it is important to avoid side reactions that yield nonreactive chain ends. It is also generally necessary to remove the condensation by-product.

(6) 实验完毕后,将黏度计从水浴中移出,清空黏度计,用纯净的溶剂冲洗。倒出废液并用纯净的氮气吹干黏度计。

思考题

(1) 画一个包含 C, η_r, η_{sp}, η_{sp}/C 和 $\ln\eta_r/C$ 值的表,并利用此表作图。记录溶剂的黏度值,Huggins 常数 k',Kraemer 常数 k'' 和测量温度。

(2) 利用马克 - 霍温克方程计算聚合物的平均分子量(在文献中查找值 K 与 α)。

实验十四　缩聚法制备尼龙 66

一、实验目的

(1) 合成尼龙 66,并计算其产率。
(2) 掌握尼龙合成过程中的涉及的基本概念与理论。
(3) 理解缩聚反应的特点。

二、实验原理

从广义上来讲,聚合反应可以被分为逐步聚合和连锁聚合反应。在逐步聚合反应中,聚合物链以缓慢、逐步的方式进行增长,然而在连锁聚合中,反应聚合物链快速增长到高分子量链。缩聚和一些开环聚合是逐步聚合反应的例子,而自由基和紫外引发聚合是连锁聚合反应的例子。

缩聚反应涉及到小分子的损失,例如水。图 2.14 是一般的逐步聚合反应的例子。

对于缩聚反应的发生,每个反应分子必须至少有两个活性基团(官能度至少为 2),官能度大于 2 时,将会产生分支和交联结构。为了得到高分子量的聚合物,通常必须用成化学计量比例的反应基团去引发反应。反应混合物应该除去带有相似官能团的杂质,避免可以生成无活性的链终端的副反应也是很重

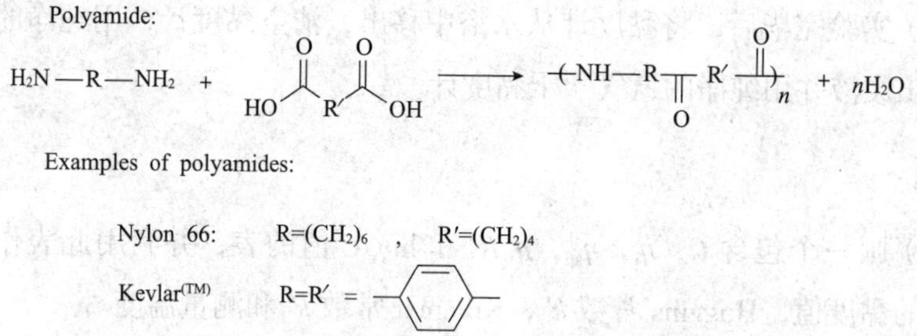

Polyamide:

$$H_2N-R-NH_2 + HOOC-R'-COOH \longrightarrow {-}[NH-R-NH-CO-R'-CO]_n{-} + nH_2O$$

Examples of polyamides:

Nylon 66: $R=(CH_2)_6$, $R'=(CH_2)_4$

Kevlar(TM): $R=R'=$ —⟨C$_6$H$_4$⟩—

Figure 2.14 Schematic diagram of step-growth polymerization

Two different functional groups are involved in any condensation polymerization, which typically are two different reactive groups on two monomers (AA and BB), or on the same monomer (AB).

One of the challenges of achieving high molecular weight in condensation polymerization is to maintain exact stoichiometry of reactants. To avoid this issue, a Nylon salt may be formed in a first stage. The salt is then polymerized in a melt reaction at temperatures of 200 - 275 ℃ and pressures of up to 20 atmospheres. Molten fiber may be drawn into high strength fibers. Additionally, nylon is used as an engineering thermoplastic and processed via extrusion, injection molding, blow molding, and other common plastic processing techniques for applications such as gears and bearings, pump bodies and brushes. Nylon polymers are characterized by high melting points and high crystallinity which provide high mechanical strength, rigidity, toughness and chemical resistance. Limitations of nylon include moisture absorption and sensitivity to degradation on environmental exposure.

Nylon 66 is produced here by the melt method. We will first produce "nylon salt" from the reaction of hexamethylene diamine (HMDA) and adipic acid. In this particular case, the stoichiometric equivalence of the functional groups is achieved by isolating the 1:1 salt before allowing the condensation to take place. Then the nylon salt is converted, under pressure and heat, to nylon 66.

3. Instruments and reagents

Instruments: Nitrogen cylinder; beakers (250 mL, 600 mL); 100 mL graduated cylinder; filter paper; Buchner's funnel; silicon oil; ring stand; ring; wire mesh; thermometer (-20 ℃ to 350 ℃); Bunsen burner; large ignition tube; stopper to fit ignition tube; heat resistant tape; wire mesh screen; one hole stopper (to fit ignition tube) with tubing to attach to vacuum pump; tongs; vacuum pump; electric balance; (optional) melting point apparatus.

Reagents: Adipic acid ($HOOC(CH_2)_4COOH$); hexamethylenediamine (HMDA; $H_2N(CH_2)_6NH_2$); ethanol.

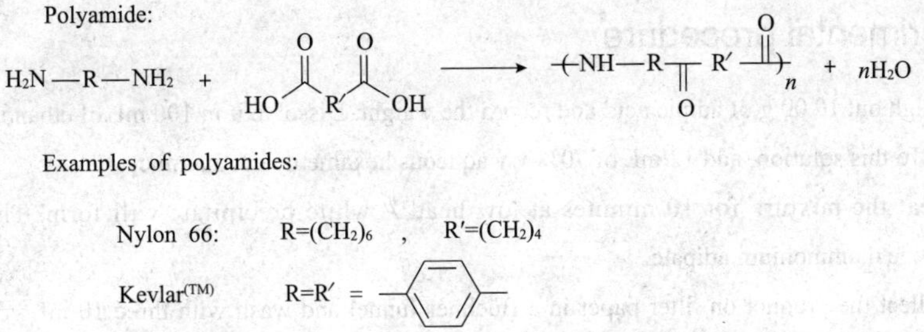

Polyamide:

$$H_2N-R-NH_2 + HOOC-R'-COOH \longrightarrow \mathrm{-[NH-R-CONH-R'-CO]}_n\mathrm{-} + nH_2O$$

Examples of polyamides:

Nylon 66: $R=(CH_2)_6$, $R'=(CH_2)_4$

Kevlar(TM) $R=R' = $ ⟨对亚苯基⟩

图 2.14　逐步聚合反应示意图

要的。通常我们必须要移除浓缩的副产物。任何缩聚反应都需要涉及两种不同的官能团，典型的例子是两种不同的活性官能团在两种单体上（AA 和 BB）和同一单体上（AB）。

在缩聚反应中，实现高分子量面临的挑战之一是保持反应物精确的化学计量比。为了避免这一问题，首先在第一阶段形成尼龙盐，尼龙盐在 200~275 ℃、2 MPa 条件下在熔化反应中进行聚合。熔化的纤维可以被拉成高强度纤维。除此之外，尼龙被用作一种工程热塑性材料，可以通过挤出、注射成型、吹塑成型和其他常见的塑料加工技术进行加工来制备一些应用型产品，例如齿轮、轴承、泵体和刷子。尼龙产品具有高熔点、高结晶度（这会提供高的机械强度、硬度、韧性和耐化学性）的特性。尼龙的局限性包括吸湿性、暴露在环境中易降解等。

本实验通过熔化法制备尼龙 66，首先通过己二胺（HMDA）和己二酸反应制备尼龙盐，在这种情况下，在缩聚发生之前，通过隔离 1∶1 盐来实现官能团的化学计量平衡。然后，在一定压力和温度下尼龙盐转化成尼龙 66。

三、试剂与仪器

仪器：氮气瓶；烧杯（250 mL、600 mL）；100 mL 量筒；滤纸；布氏漏斗；硅油；环架；环；金属线网；温度计（-20~350 ℃）；煤气喷灯；大型点火管；点火管配套塞；防热胶带；网筛；连接真空泵的点火管单孔塞；钳子；真空泵；分析天平；熔点仪（可选）。

试剂：己二酸（HOOC（CH$_2$）$_4$COOH）；己二胺（H$_2$N（CH$_2$）$_6$NH$_2$，简称 HMDA）；乙醇。

4. Experimental procedure

(1) Weigh out 10.00 g of adipic acid and record the weight. Dissolve it in 100 mL of ethanol in a 250 mL beaker. To this solution, add 12 mL of 70%-v/v aqueous hexamethylenediamine.

(2) Heat the mixture for 10 minutes at low heat. A white precipitate will form. This is the hexamethylene diammonium adipate.

(3) Collect the product on filter paper in a Büchner funnel and wash with three 10 mL volumes of ethanol. Air dry the product until no ethanol remains. Weigh the dry product and calculate the %-yield by mass.

(4) Use 300 mL of silicon oil in a 600 mL beaker to set up an oil bath. Place on the ring stand with ring and wire gauze over the Bunsen burner in the hood. Preheat the oil bath to 215 ℃ .❶

(5) Place 10 g of the salt in an ignition tube, purge with nitrogen, and seal with a rubber stopper. Wrap heat resistant tape around top of ignition tube and rubber stopper to seal. Place a wire mesh screen around the ignition tube.

(6) Place the ignition tube set-up in the 215 ℃ oil bath and leave for one hour. Monitor the oil bath temperature and adjust the burner as needed to keep the temperature as close as possible to 215 ℃ .

(7) Carefully remove the ignition tube from the oil bath using tongs and allow it to cool to near room temperature. Raise the temperature of the oil bath to 265–270 ℃ .

(8) Once the ignition tube is cool, remove the wire screen, unwrap the tape, and remove the stopper. Add an additional 0.55 g of adipic acid to the ignition tube. Insert a one hole stopper with stem into the ignition tube and attach the stem to one of the vacuum pump inlet manifold hoses. Turn on the vacuum pump and place the ignition tube in the 267 ℃ oil bath for one hour.

(9) After one hour, turn off the burner but keep the vacuum on. Allow the ignition tube to cool and then remove it from the oil bath. Turn off the vacuum. Remove the nylon 66 plug carefully from the ignition tube.

(10) Submit the sample to your instructor along with your data and the answers to the questions.❷

Questions

(1) Where is nylon 66 used as a final product of this experiment?

(2) What properties might be used to characterize this polymer?

(3) What condensate is removed during the reaction?

---

❶ Students should understand the dangers associated with the hot silicon oil, the open flame, and the vacuum apparatus.

❷ Pour the solvents from this experiment into an organic waste container.

四、实验步骤

（1）称取 10.00 g 己二酸，并记下质量值，将其加入装有 100 mL 甲醇的 250 mL 烧杯中。充分相溶后，向上述溶液中加入体积分数为 70% 的己二胺水溶液。

（2）在较低温度下加热上述混合物 10 min，将会产生己二酸己二胺盐的白色沉淀，即为"尼龙盐"。

（3）通过减压抽滤在滤纸上收集产物，用 10 mL 的甲醇洗三次，风干产物，直到没有甲醇残留。称风干后产物的重量，计算质量产率。

（4）将 300 mL 硅油加入到 600 mL 烧杯中作为油浴锅，且将其放在煤气喷灯上面的由环和金属线网构成的环架上。预加热油浴锅到 215℃。❶

（5）在一个点火管里放 10 g 尼龙盐，N_2 吹扫，然后用橡胶塞密封，在点火管和橡胶塞的顶部周围包裹防热胶带，然后在点火管周围放置一个网筛。

（6）把上述点火管置于 215 ℃ 的油浴锅中 1 h，监控油浴锅温度、调节火炉来保持温度尽可能地接近 215℃。

（7）用钳子小心地将点火管从油浴中移出，冷却至室温。然后将油浴锅的温度升至 265~270℃。

（8）待点火管冷却后，移去网筛、去掉包裹的胶带、移除塞子，向点火管中再加入 0.55 g 己二酸。然后在点火管上塞入单孔塞，此单孔塞接有连接到真空泵的软管。打开真空泵，将点火管放置在 267 ℃ 的油浴锅中 1 h。

（9）1 h 后，关闭火炉，保持真空泵开着，在此条件下，冷却点火管，然后将点火管从油浴中移出，关闭真空泵。最后从点火管中小心的取出尼龙 66 产品。

（10）实验数据、实验问题的答案和最终的尼龙 66 产品交给老师。❷

思考题

（1）本实验的最终产品——尼龙 66 被用在哪些地方。

（2）合成的这种高分子——尼龙 66 的突出特性是什么。

（3）在反应过程中，什么浓缩物被移除了。

---

❶ 提醒学生应该意识到高温热硅油、开放火焰和真空设备的危险性，并注意防护。

❷ 本实验反应废溶剂要倒入有机废物桶中。

(4) Is the synthesis of nylon 66 a chain-growth polymerization or is it a step-growth polymerization? Discuss these two types of polymerization.

Experiment 15 Kinetics of bulk polymerization of methyl methacrylate

1. The purposes

(1) Master the principle and method of measuring polymerization rate by dilatometer.
(2) To understand the mechanism of bulk polymerization of methyl methacrylate.
(3) Understand the experimental data processing and calculation methods.

2. Experimental principle

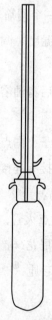

Figure 2.15 Schematic diagram of dilatometer

There are two kinds of methods to measure the rate of polymerization: direct method and indirect method. The direct methods include chemical analysis, evaporation and precipitation. The most commonly used method is precipitation method. That is, in the polymerization process, the polymer is regularly sampled and precipitated with precipitant. Separate and refine the precipated polymer, then dry and weigh to obtain the amount of polymer. The indirect method is to determine the specific volume, viscosity, refractive index, dielectric constant, absorption spectrum and other physical properties changes in the polymerization process, and indirectly calculate the amount of polymer.

The principle of measuring polymerization rate by dilatometer is to use the linear relationship between volume shrinkage and conversion during polymerization. The dilatometer is a special polymerizer with a capillary in the upper part, with a piston grinding mouth and a hook in the middle. As shown in Figure 2.15, the volume change of the reaction system can be read directly from the drop of the capillary liquid level. The conversion rate is calculated as follows:

$$C = \frac{V'}{V} \times 100\% \tag{1}$$

Where, C is the conversion rate. V' is the shrinkage number of the system at different reaction time t, which is read out from the capillary scale of the dilatometer, V represents the volume shrinkage number of monomer converted to polymer at 100% of the capacity.

$$V = V_M - V_P = V_M \left(1 - \frac{d_M}{d_P}\right) \tag{2}$$

(4)尼龙66的合成属于连锁聚合还是逐步聚合,讨论这两种类型的聚合。

实验十五　甲基丙烯酸甲酯本体聚合动力学

一、实验目的

(1)掌握膨胀计测定聚合反应速率的原理和方法。
(2)了解甲基丙烯酸甲酯本体聚合的机理。
(3)了解实验数据处理和计算方法。

二、实验原理

聚合反应速率的测定方法有直接法和间接法两类。直接法有化学分析法、蒸发法、沉淀法。最常用的是沉淀法,即在聚合过程中定期取样,用沉淀剂使聚合物沉淀,然后分离、精制、干燥、称重,求得聚合物量。间接法是测定聚合过程中比容、黏度、折光率、介电常数、吸收光谱等物性的变化,间接求其聚合物的量。

膨胀计法测反应速率的原理是利用聚合过程中体积收缩与转化率呈线性关系。膨胀计是上部装有毛细管的特殊聚合器,中间有活塞磨口和挂钩,如图2.15所示,反应体系的体积变化可直接从毛细管液面下降读出。根据下式计算转化率:

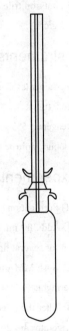

图2.15　膨胀计示意图

$$C = \frac{V'}{V} \times 100\% \tag{1}$$

式中,C为转化率;V'表示不同反应时间t时体系收缩数,从膨胀计的毛细管刻度读出;V表示该容量下单体100%转化为聚合物时的体积收缩数。

$$V = V_M - V_P = V_M \left(1 - \frac{d_M}{d_P}\right) \tag{2}$$

Where, d is the density, and the subscripts M and P denote monomer and polymer respectively.

The polymerization rate can be expressed by the consumption of monomer or the production of polymer in unit time. At low conversion, the reaction rate R can be obtained by the following formula.

$$R = \frac{d[M]}{dt} = M\frac{dc}{dt} k[I]^{1/2} M \qquad (3)$$

From the integral of the above formula (3), we can get the change rule of conversion rate with time, formula (4). Plot $\ln \frac{1}{1-C}$ for t with the slope of $k[I]^{1/2}$. At low conversion, [I] can be considered as constant, that is, equal to the initial concentration of initiator $[I]_0$. The total rate constant k can be obtained.

$$\ln \frac{[M]_0}{[M]} = \ln \frac{1}{1-C} = k[I]^{1/2} t \qquad (4)$$

In this experiment, the bulk polymerization of methyl methacrylate (MMA) was initiated by azodiisobutyronitrile (AIBN) at 60 ℃. The density of MMA at 60 ℃ is 0.8957 g/cm³, and that of PMMA is 1.179 g/cm³.

3. Instruments and reagents

Instruments: Constant temperature water bath; dilatometer; separating funnel; measuring cylinder; glass rod; timer.

Reagents: 5% sodium hydroxide solution; anhydrous sodium sulfate; methyl methacrylate (MMA); azodiisobutyronitrile (AIBN); acetone; distilled water.

4. Experimental procedure

(1) Purification of methyl methacrylate (MMA)

① Add 300 mL of MMA into a 500 mL separating funnel, wash MMA with 50 mL of 5% NaOH solution for several times until the water layer changes from red to colorless;

② Wash MMA to neutral with 50 mL of deionized water, and then add anhydrous sodium sulfate to dry overnight;

③ Filter the dried MMA and conduct vacuum distillation directly, collect the fraction and put it into the refrigerator at -20 ℃ for reserve.

Note: the refined MMA should not be placed for a long time because it is easy to self polymerize after removing the polymerization inhibitor.

(2) Kinetics of polymerization experiment

Accurately measure 20 mL of MMA and 0.1888 g of AIBN, mix them evenly in a 50 mL beaker. Then pour them into the lower part of the dilatometer to the semi grinding mouth, and insert the capillary tube. At this time, the liquid level rises to the scale of capillary (1/4 ~ 1/3) (0.2 mL is taken in this experiment, i.e. two grids). After checking whether there are bubbles in the dilatometer, fix the capillary

式中，d 为密度，下标 M、P 分别表示单体和聚合物。

聚合速率可以用单位时间内单体的消耗量或者聚合物的生成量来表示，在低转化率下，反应速率 R 可用式（3）得。

$$R = \frac{d[M]}{dt} = M\frac{dc}{dt} k[I]^{1/2} M \qquad (3)$$

由式（3）积分得到式（4），可以得到转化率随时间的变化规律，以 $\ln\frac{1}{1-C}$ 对 t 作图，其斜率为 $K[I]^{1/2}$。在低转化率下，[I] 可认为不变，即 [I] 等于引发剂起始浓度 $[I]_0$。则可得反应总速率常数 K。

$$\ln\frac{[M]_0}{[M]} = \ln\frac{1}{1-C} = k[I]^{1/2} t \qquad (4)$$

本实验以偶氮二异丁腈（AIBN）为引发剂引发甲基丙烯酸甲酯（MMA）在 60 ℃下聚合。甲基丙烯酸甲酯 MMA 在 60℃的密度取 $d_M^{60}=0.8957$ g/cm³，聚甲基丙烯酸甲酯（PMMA）取 $d_P^{60}=1.179$ g/cm³。

三、实验仪器

仪器：恒温水浴槽；膨胀计；分液漏斗；量筒；玻璃棒；秒表。

药品：5% 氢氧化钠溶液；无水硫酸钠；甲基丙烯酸甲酯（MMA）；偶氮二异丁腈（AIBN）；丙酮；蒸馏水。

四、实验步骤

（1）甲基丙烯酸甲酯（MMA）的精制

①在 500 mL 分液漏斗中加入 300 mL MMA，用 50 mL 的 5% NaOH 溶液多次洗涤 MMA，直至水层由红色变为无色；②用 50 mL 去离子水将 MMA 洗涤至中性，然后加入无水硫酸钠干燥过夜；③将干燥后的 MMA 过滤后直接进行减压蒸馏，收集馏分后放入 −20℃冰箱保存备用。注意：精制后的 MMA 不宜长久放置，因为去除阻聚剂后容易自聚。

（2）聚合动力学实验

准确量取 20 mL MMA 和 0.1888 g AIBN，在 50 mL 烧杯内混合均匀后，倒入膨胀计下部至半磨口处，插入毛细管，此时液面上升至毛细管（1/4~1/3）刻度处（本实验取上升 0.2 mL，即两大格），检查膨胀计内有无气泡后，用橡皮

tube and the lower part of the dilatometer with rubber bands❶.

Place the dilatometer fixed with rubber bands into a constant temperature water bath at 60 ℃ (note that the liquid level in the dilatometer is always lower than that in the water bath). Because of the thermal expansion, the liquid level in the capillary rises continuously. When the liquid level is stable, it can be considered that the system has reached the thermal equilibrium. Record the time and the liquid level height of the dilatometer as the starting point of the experiment, and observe the change of the liquid level. The period from the rise of the liquid in the tube to the beginning of its decline is called the induction period. When the liquid level begins to drop, it means that the reaction begins. Then, read the volume change of the capillary every 5 minutes until the end of the experiment. Suggest recording six time points in this experiment. Too many experimental points will cause the viscosity of the system to be too high and make it difficult to remove the capillary.

After the experiment, the reaction bottle should be separated from the capillary tube immediately to prevent the dilatometer from sticking. Wash the capillary tube and the reaction bottle with acetone. Do not pour MMA or other toxic reagents into the sewer; do not directly pour the waste liquid from washing glassware containing MMA or other toxic reagents into the water tank.❷

5. Experimental data processing

Table 2.6 Record of experimental data

t/min	7	12	17	22	27	32
V_t/mL						
V'/mL						
C						
$\ln 1/(1-C)$						

① $C = \dfrac{V'}{V} \times 100\%$, $V = V_M - V_P = V_M(1 - \dfrac{d_M}{d_P})$

② Conversion time (C-T) curve: According to formula, the conversion C under different reaction time t can be obtained. $C \sim t$ curve is obtained by plotting C to t. The reaction rate $R = [M] \dfrac{dc}{dt}$ is obtained from the slope of the curve, mol/L · min.

Questions

(1) Please analyze the cause of induction period of polymerization.

(2) Try to analyze the causes of experimental errors in this experiment.

---

❶ The monomer and initiator should be mixed evenly first, and there should be no bubble in the dilatometer when adding the reaction liquid.

❷ In this experiment, the waste solvent should be poured into the organic waste bucket.

筋固定膨胀计的毛细管与下部❶。

将橡皮筋固定好的膨胀计置入 60 ℃的恒温水浴槽中（注意膨胀计内液面始终低于水浴液面）。由于热膨胀，毛细管内液面不断上升，当液面稳定不动时，可认为体系达到热平衡，记录时间及膨胀计的液面高度作为实验起点，观察液面变化。从管内液体上升至最高点到开始下降的这段时间称为诱导期。液面一开始下降表示反应开始，计时。随后，每隔 5 min 左右读一次毛细管体积变化至实验结束，本次实验做 6 个点，点数太多，体系黏度过大，使毛细管难以取下。

实验结束后马上将反应瓶与毛细管分离，防止膨胀计粘结，用丙酮冲洗毛细管和反应瓶。不要把 MMA 或是其他有毒试剂倒入下水道；不要将冲洗含有 MMA 或是其他有毒试剂玻璃器皿的废液直接倒入水槽❷。

五、实验数据处理

表 2.6　实验数据记录

t/min	7	12	17	22	27	32
V_t/mL						
V'/mL						
C						
$\ln 1/(1-C)$						

注：① $C = \dfrac{V'}{V} \times 100\%$，$V = V_M - V_P = V_M(1 - \dfrac{d_M}{d_P})$。

② 转化率－时间（C-t）曲线：根据上式求得不同反应时间 t 下的转化率 C。以 C 对 t 作图得 C~t 曲线。

从斜率求得反应速率 $R = [M] \dfrac{dc}{dt}$，mol/L·min。

思考题

（1）分析在实验中诱导期产生的原因。

（2）试分析本实验中的实验误差产生原因。

---

❶ 单体和引发剂先混合均匀，加入反应液时要保证膨胀计内不能有气泡。

❷ 本实验反应废溶剂要倒入有机废物桶中。

Experiment 16 Syntgesis of Nylon by interfacial polymerization

1. The purposes

(1) Learn the fundamental laboratory skills necessary for making nylons.

(2) Understand the principle of interfacial polymerization.

2. Experimental principle

Nylon is the common name of polyamide, which is the general name of thermoplastic resin with repeated amide groups in the main chain. According to the chemical structure, it can be divided into aliphatic nylon, aliphatic aromatic nylon and aromatic nylon. Among them, there are many varieties of aliphatic nylon, with large output and wide application. The main varieties are nylon 6 and nylon 66. Nylons are some of the most important fibers produced commercially, but nylon can be more than just fibers. It's also used for self-lubricating gears and bearings. Nylon clay composites can also be used as under-hood automobile parts.

Interfacial polymerization is a special way of polycondensation. In interfacial polymerization, reaction proceeds at the interface of two immiscible liquid phases. Each phase contains one monomer and no strict requirement for monomer purity. The product of polymerization is insoluble in solvent and precipitates at the interface. The polymerization rate is determined by the diffusion rate of reactants to the interface. Compared with most polycondensation reactions, interfacial polymerization can get high molecular weight products faster. High molecular weight products can be obtained at room temperature without strict control of the stoichiometric number of reactants.

In this experiment, we take the preparation of nylon 610 as an example to learn the principle of interfacial polymerization. Making nylon 610 is even easier if diacyl chloride is used to replace diacid acid and react with diamine to prepare nylon 610. This is because acyl chlorides are much more reactive than acids. The reaction was carried out in a two-phase system, in which sebacoyl chloride existed in the organic phase and hexanediamine in the aqueous phase. Put the two solutions into in the same beaker. Because the two solutions could not be mutually soluble, there are two phases in the beaker. At the interface of the two phases, diacylchloride and diamine contact each other and polymerize. The reaction formula is shown in Figure 2.16. However, this method is rarely used in industrial production dut to the following reasons: 1) the price of acyl chloride is much higher than that of acid; 2) acyl chloride has a great odor and is more toxic than acid; 3) the consumption of reaction solvent is large; 4) the utilization

$$n\text{ClOC(CH}_2)_8\text{COCl} + n\text{NH}_2(\text{CH}_2)_6\text{NH}_2 \longrightarrow \text{\{NH(CH}_2)_6\text{NH}-\text{OC(CH}_2)_8\text{CO\}}_n + 2n\text{HCl}$$

Figure 2.16 Reaction route of nylon-610

实验十六 尼龙的界面聚合

一、实验目的

（1）学会制备尼龙过程中的基本实验技能与方法。

（2）理解界面聚合的基本原理。

二、实验原理

尼龙是聚酰胺的俗称，是分子主链上含有重复酰胺基团的热塑性树脂的总称。按化学结构可分为脂肪族尼龙、脂肪-芳香族尼龙以及芳香族尼龙。其中脂肪族尼龙品种多，产量大，应用广泛，主要品种是尼龙6和尼龙66。尼龙是最重要的商业生产纤维中的一部分，但尼龙不仅仅是纤维，也可被用作自润滑齿轮和轴承。尼龙-黏土复合材料还可被用作汽车零件。

界面聚合是缩聚反应特有实施方式。在界面聚合中，反应在两种互不相溶液相的界面发生，每一种液相中包含一种单体。聚合反应产物不溶于溶剂在界面析出。聚合速度由反应物扩散到界面的速度决定。对单体的纯度要求不严，与大部分缩聚反应相比，界面聚合可更快地得到高分子量的产物。高分子量产物在室温下即可获得，不需要严格的控制反应物的化学计量数。

本实验以尼龙610的制备为例来学习界面聚合原理。如果采用二酰氯代替二酸与二胺反应制备尼龙610，由于酰氯比酸更活泼，这样的制备会更简单。该反应在两相体系中进行，癸二酰氯存在于有机相，己二胺存在水相中，然后将这两种溶液放置于同一烧杯中，两种溶液不能互溶，故在烧杯中存在两相。在两相的界面处，二酰氯和二胺彼此接触，将会发生聚合，反应式如图2.16所示。然而这种方法很少用在工业生产上，因为酰氯比酸的价格要高很多，酰氯有很大的臭味且比酸的毒性更大，反应溶剂消耗量大，设备利用率低，并且这

$$n\text{ClOC(CH}_2)_8\text{COCl} + n\text{NH}_2(\text{CH}_2)_6\text{NH}_2 \longrightarrow +\!\!\!\left[\text{NH(CH}_2)_6\text{NH}-\text{OC(CH}_2)_8\text{CO}\right]_n\!\!\!+ 2n\text{HCl}$$

图2.16 生成尼龙-610的反应路线

rate of equipment is low; 5) the strength of fiber prepared by this method is not large enough.

3. Instruments and reagents

Instruments: Electric balance; vaccum oven; beakers (100 mL × 2); glass stirring rod; forceps; a test tube (25 mm×150 mm); metal spatula; paper towel.

Reagents: hexamethylene diamine; sebacoyl chloride; hexane; toluene; acetone; methanol.

4. Experimental procedure

(1) Dissolve about 1 g of hexamethylene diamine in 25 mL of water in a 100 mL beaker.

(2) Make solution of about 1 g of sebacoyl chloride in 25 mL hexane.

(3) Gently pour the sebacoyl chloride/hexane solution on top of the hexamethylene diamine/water solution in the beaker, using a glass rod to pour down. Allow the reaction mixture to stand undisturbed for about 1 minute as a white film of Nylon polymer forms at the liquid interface ❶❷.

(4) Reach into the center of the beaker with forceps, and grasp the Nylon film at the interface. Slowly pull out the polymer line from the vertical direction of the solution, and wrap around a large tube (25mm × 150 mm) from the top. Carefully pull out the nylon thread from the center of the beaker at a constant speed and rotate it around the test tube. If the nylon thread breaks, a new nylon thread can be produced with forceps again. However, if possible, we should pull out the unbroken nylon thread continuously until most of the reaction is completed ❸.

(5) Rinse the excess solvent from the surface of the Nylon spool by rotating the test tube under a stream of tap water. Slide the Nylon from the test tube as an intact loop by inserting a metal spatula between the polymer and the glass surface. Observe the physical appearance and texture of the Nylon. Lay the polymer on a paper towel to drain.

(6) Remove a small amount of the polymer to test its solubility in toluene, acetone and methanol.

(7) The samples were dried in a vacuum oven at 50 ℃, and the prepared polymer was weighed to calculate the yield.

Questions

(1) What would be the effect of adding a monofunctional monomer to the reaction mixture?

(2) What are the disadvantages of interfacial polymerization in comparison to bulk polymerization?

---

❶ Do not pour any chemicals from this experiment down the sink! In cleaning the reaction beaker and test tube, residual deposits of Nylon fiber may be removed satisfactorily by scraping with a metal spatula.

❷ In order to improve the reaction efficiency, stirring can be used to increase the total interfacial area.

❸ Acid substance is produced in the reaction process, which needs to be neutralized by adding an appropriate amount of alkali into the system.

种方法制得纤维强度不够大。

三、仪器和试剂

仪器：电子天平；真空烘箱；烧杯（100 mL×2）；玻璃棒；镊子；试管（25mm×150 mm）；金属铲；纸巾。

试剂：己二胺；癸二酰氯；己烷；甲苯；丙酮；甲醇。

四、实验步骤

（1）将1 g己二胺加入到装有25 mL水的100 mL烧杯中，使其溶解；

（2）将1 g癸二酰氯溶解于25 mL己烷中，制备成溶液；

（3）将癸二酰氯/己烷溶液轻轻倒入己二胺/水溶液上方，用玻璃棒引流。反应混合物静置1min，在液相界面形成了尼龙聚合物的白色薄膜❶❷；

（4）用镊子接触烧杯的界面中心，抓住界面的尼龙薄膜，慢慢地从垂直溶液方向拔出聚合物线，围绕一个大试管（25 mm×150 mm）从顶端开始缠绕。以恒定的速度从烧杯的中心小心地拉出尼龙线并旋转使其缠绕在试管周围，如果这条尼龙线断裂，用镊子可产生新的尼龙线。如果可能的话，应该连续拉出未破损的尼龙线，直到大部分反应完成❸。

（5）在一股水龙头水流的冲洗下，旋转试管，使尼龙线的表面的过量溶剂被冲洗掉。在聚合物和玻璃表面插入一个金属铲，使尼龙以完整线环的形式从试管上滑动。观察尼龙的物理性质和质地，将制得的聚合物放在纸巾上吸干。

（6）取少量的聚合物测试其在甲苯、丙酮、甲醇中的溶解性。

（7）将样品在50℃真空烘箱中干燥，称重制备的聚合物，计算产率。

思考题

（1）向反应混合物中加单官能度单体将会有什么影响？

（2）比较本体聚合，界面聚合有哪些缺点？

---

❶ 本实验的任何化学废液不要倒入水槽中。在清洗反应烧杯和试管时，尼龙纤维的残余物可以通过金属铲刮抹来有效地去除。

❷ 为了提高反应效率，可以采用搅拌的方法提高界面总面积。

❸ 反应过程有酸性物质生成，可以在体系中加入适量的碱中和以提高反应速率。

Experiment 17　Fabrication of self-healing flexible transparent conductive film

1. The purposes

(1) Learn the fundamental laboratory skills and methods for synthesizing the self-healing polyurethane.

(2) Learn how to prepare copper nanowires via hydrothermal method.

(3) Understand thermal reversible reaction principle of self repairing polyurethanethe.

(4) Learn how to characterize the performance of self-healing flexible transparent conducting film.

2. Experimental principle

With the development of science and technology and continuous improvement of the people's living standard, transparent conductors in the future should be bendable, expandable, and deformable into complex shapes with good reliability and integration, such as in the use of stretchable display, energy-related devices, stretchable solar cells, stretchable integrated circuits, and elastomeric organic transistors. Though indium tin oxide (ITO) has been mainly applied in the field of transparent conductors, several critical issues, such as high cost and scarcity of raw materials, and brittleness limited its wide use in many fields.

In this experiment, copper nanowires (CuNWs) are prepared by hydrothermal method. Self-healing polyurethane can be synthesized with a two-step conventional method. Firstly, the polyurethane pre-polymer with –NCO group was prepared using 2, 4-tolulene diisocyanate and PEG 600. Then, the polyurethane pre-polymer containing furan rings (PU-Furan) was synthesized by capping polyurethane pre-polymer with furan methylamine. Finally, PU-Furan was cross-linked with N, N'-(4,4'-methylenediphenyl) dimaleimide via a Diels-Alder reaction to form the healable network polymer (DA-PU). The reaction route is shown in Figure 2.17. A schematic fabrication procedure of a healable nanocomposite conductor based on CuNWs and self-healing polyurethane can be found in Figure 2.18. Finally, the performances of the self-healing flexible transparent conducting film are characterized by polarizing microscopy, UV-Vis spectrophotometer, 4-point probe sheet resistance meter and one simple light-emitting diode (LED) circuit set.

实验十七　自修复柔性透明导电薄膜的制备

一、实验目的
（1）学会制备自修复聚氨酯过程中的基本实验技能与方法。
（2）学会利用水热法制备铜纳米线的方法。
（3）理解自修复聚氨酯热可逆反应原理。
（4）了解自修复柔性透明导电薄膜的性能表征及应用。

二、实验原理

随着科学技术的发展和人民生活水平的不断提高，具有柔性特征的平面显示器、能源相关设备、太阳能电池、集成电路和有机晶体管等的应用将越来越广。为了更好地适应各种复杂情况，可弯曲变形的透明导电薄膜需求越来越大。ITO（Indium Tin Oxide）作为一种在光电子器件应用广泛的导电材料，具有透光性好、电阻率低、易刻蚀和易低温制备等优点。但由于铟矿稀少，ITO薄膜制备工艺复杂，所以该薄膜价格较高。同时，该薄膜较脆，使用过程中容易产生裂纹，限制了其在柔性基底上的应用。

本实验中铜纳米线（CuNWs）是通过常规水热法合成。自修复聚氨酯的合成分为两步，首先是通过2,4-甲苯二异氰酸酯和聚乙二醇600合成两端为-NCO官能团的聚氨酯预聚体，并用呋喃甲胺进行封端处理，制备出以呋喃环为端基的聚氨酯（PU-Furan）。然后将PU-Furan与4,4'-双马来酰亚胺基二苯甲烷反应，在聚氨酯基材中引入热可逆Diels-Alder键，从而实现自修复功能（DA-PU）。反应方程式如图2.17所示。图2.17（a）为呋喃环封端聚氨酯图2.17为（b）为自修复聚氨酯合成路线图。自修复柔性透明导电薄膜的制备示意图如图2.18所示。最后通过偏光显微镜、紫外-可见分光光度计、四探针测试仪以及一组简易的LED电路对自修复导电薄膜的性能进行了表征。

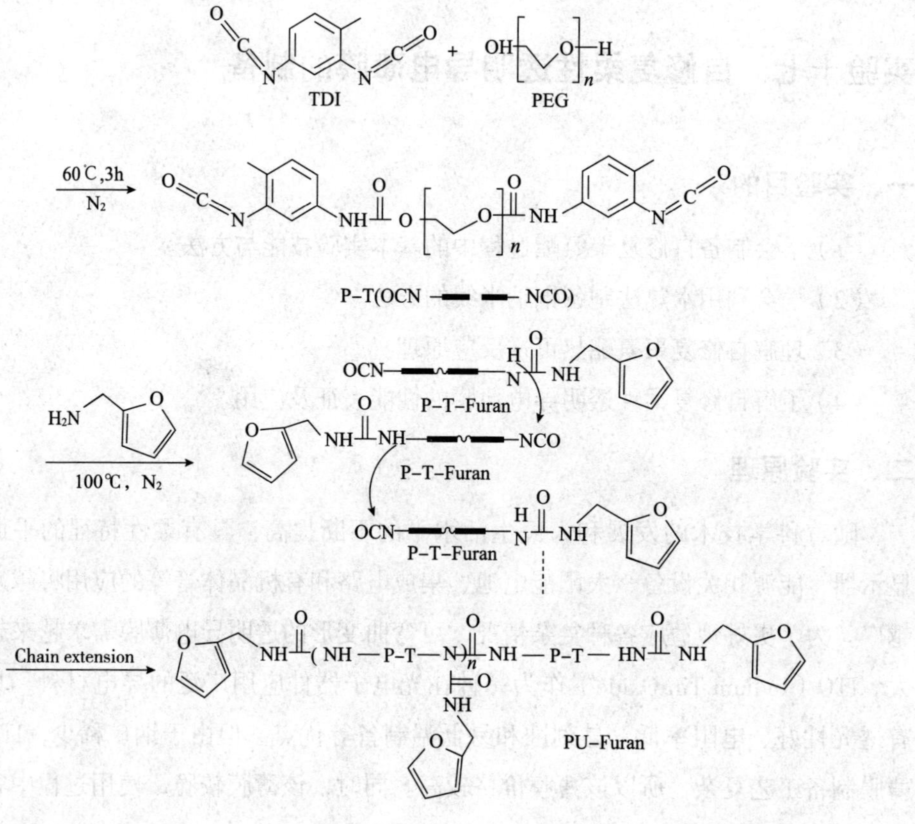

(a)

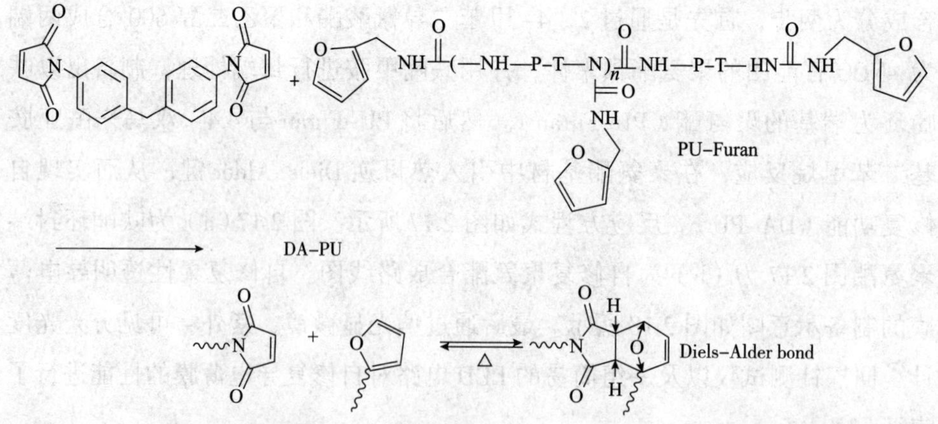

(b)

Figure 2.17 A reaction chart for the synthesis of a self-healing polyurethane

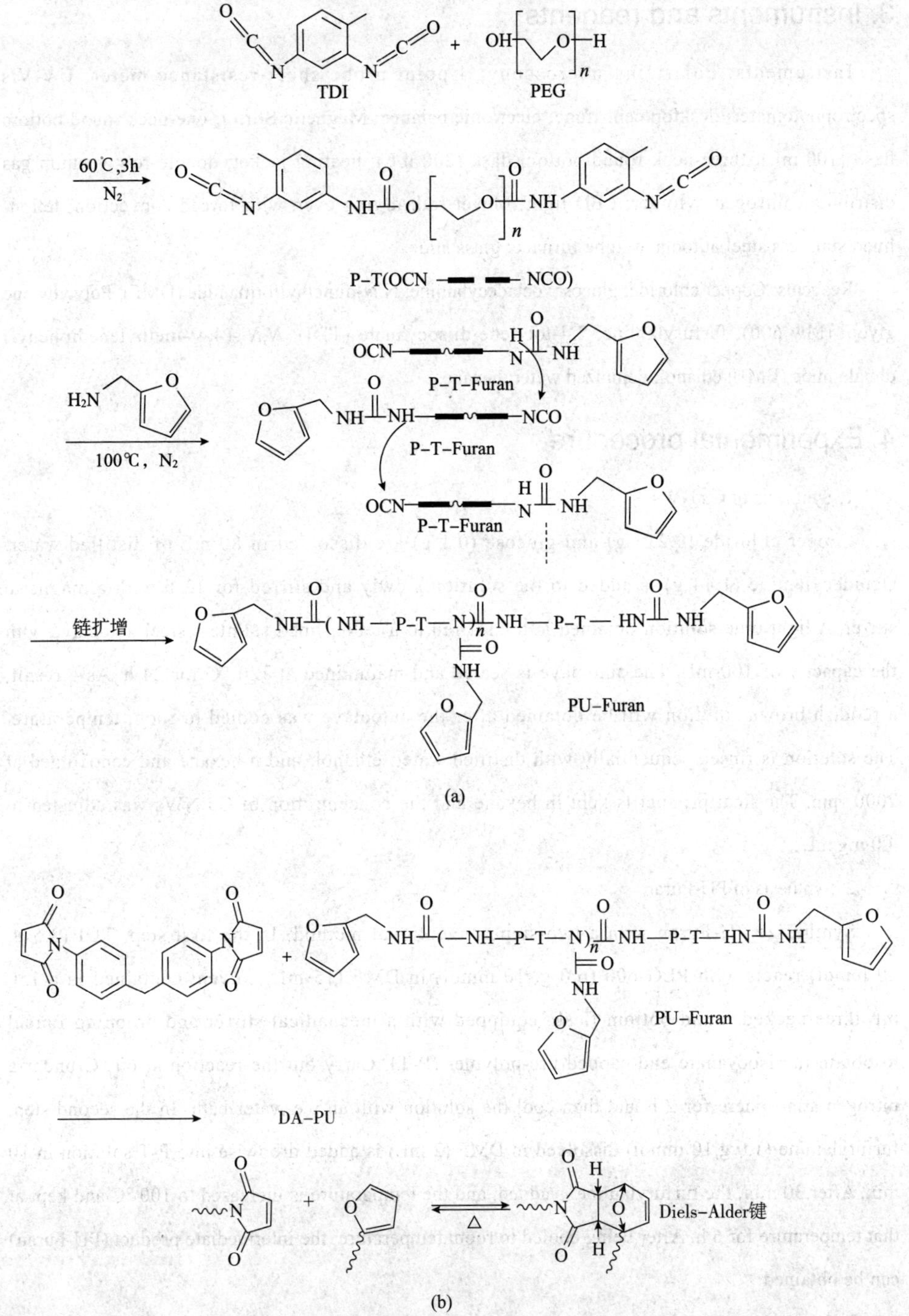

图 2.17 自修复聚氨酯合成路线图

3. Instruments and reagents

Instruments: Polarizing microscopy; 4-point probe sheet resistance meter; UV-Vis spectrophotometer; desktop centrifuge; electronic balance; Magnetic Stirrer; one-neck round bottom flask (100 mL); three-neck round bottom flask (200 mL); heating jacket; double-row vacuum gas distributor; nitrogen cylinder; LED light; Meyer rod; drying oven with forced convection; teflon-lined stainless steel autoclave; tube furnace; glass slide.

Reagents: Copper chloride; glucose; octadecylamine; N,N-dimethylformamide (DMF) Polyethylene glycol (Mw 600); furfurylamine; 2,4-tolylene diisocyanate (TDI); N,N'-(4,4'-methylenediphenyl) dimaleimide (BMI); ethanol; deionized water; hexane.

4. Experimental procedure

1. Synthesis of Cu NWs

Copper chloride (0.215 g) and glucose (0.1 g) are dissolved in 80 mL of distilled water. Octadecylamine (1.44 g) is added to the solution slowly and stirred for 12 h with a magnetic stirrer. A light blue solution obtained is transferred to a teflon-lined stainless steel autoclave with the capacity of 100 mL. The autoclave is sealed and maintained at 120 °C for 24 h. As a result, a reddish brown solution will be obtained after the autoclave was cooled to room temperature. The solution is rinsed sequentially with distilled water, ethanol, and n-hexane and centrifuged at 2000 rpm. The final product is kept in hexane and the concentration of Cu NWs was adjusted to 4.0 mg mL^{-1}.

2. Synthesis of PU-Furan

Synthesize PU-Furan with a two-step conventional method. In the first step, TDI (3.5 g, 20 mmol) reacts with PEG-600 (6.0 g, 10 mmol) in DMF (15 mL) solvent contained in a 150 mL three necked round bottom flask, equipped with a mechanical stirrer and dropping funnel to obtain the isocyanate end-capped pre-polymer (P-T). Carry out the reaction at 60 °C under a nitrogen atmosphere for 2 h and then cool the solution with an ice-water bath. In the second step, furfurylamine (1.0 g,10 mmol) dissolved in DMF (2 mL) is added dropwise into P-T solution in 10 min. After 30 min, the furfurylamine is added, and the temperature is increased to 100 °C and kept at that temperature for 6 h. After being cooled to room temperature, the intermediate product (PU-Furan) can be obtained.

3. Fabrication of self-healing flexible transparent conducting film

(1) Treatment of glass slide

三、仪器和试剂

仪器材料：偏光显微镜；四探针电阻仪；721型可见分光光度计；台式离心机；电子分析天平；磁力搅拌器；单口烧瓶（100mL）；三口烧瓶（200 mL）；调温加热套；双排管气体分配器；氮气瓶；市售LED灯；涂布棒；鼓风干燥箱；水热合成反应釜；载玻片；管式炉。

试剂：二水合氯化铜（$CuCl_2 \cdot 2H_2O$）；葡萄糖（$C_6H_{12}O_6$）；十八胺；无水 N，N-二甲基甲酰胺（DMF）；无水聚乙二醇（PEG-600）；2,4-甲苯二异氰酸酯（TDI）；呋喃甲胺、4,4'-双马来酰亚胺基二苯甲烷（BMI）；无水乙醇；去离子水；正己烷。

四、实验步骤

（1）铜纳米线的制备。

准确称取 0.215 g $CuCl_2$ 和 0.1 g 葡萄糖，溶于 80 mL 去离子水中并转入 100 mL 单口烧瓶，准确称取 1.44 g 十八胺，缓慢加入溶液中，并进行磁力搅拌 12 h，混合均匀后，得到浅蓝色溶液，将其倒入有特氟龙衬里的反应釜中，置于烘箱内，120℃加热反应 24 h，待其冷却至室温后，取出反应釜，即得到棕红色溶液。将棕红色的溶液依次用去离子水、乙醇和正己烷离心清洗，最后将铜纳米线保留在正己烷中，浓度为 4 mg/mL。

（2）呋喃环为端基的聚氨酯预聚体的制备。

先用氮气把装有搅拌装置、通氮气装置的三口烧瓶（150 mL）中的水除干净。再在三口烧瓶中加入 3.5 g 2,4-甲苯二异氰酸酯（TDI），10g PEG-600 和 15 mL DMF，搅拌溶解。待两者混合均匀后，在氮气氛围下保持温度 60℃加热 2 h，得到聚氨酯预聚体。

在冰水浴条件下，将溶于 DMF（2 mL）的呋喃甲胺（1.0 g）滴入充分除氧的聚氨酯预聚体中。反应 30 min 后，升温到 60℃反应 1 h，再升温到 100℃反应 4 h，最后升温到 120℃反应 4 h，制得以呋喃环为端基的聚氨酯（PU-Furan）。

（3）自修复柔性透明导电薄膜的制备。

①玻璃板的处理：将载玻片擦拭后，放入 2% 的 NaOH 溶液中，3~5 min 后，将载玻片从 NaOH 溶液中取出，放入去离子水中；3~5 min 后，用镊子取

The glass slide is first immersed in 2% NaOH for 3-5 min, then it is picking up from NaOH solution by tweezers and put into deionized water for another 3-5 min. Finally, the glass slide is immersed in acetone for 3-5 min and dried with blowing nitrogen for further use.

(2) Preparation of copper nanowire film

CuNWs dispersed in n-hexane at a concentration of 4.0 mg/mL is coated onto a glass substrate by a Meyer rod. The preformed CuNWs on the donor glass are annealed in a tube furnace under H_2 atmosphere for 1 h at 200 °C.

(3) Fabrication of self-healing flexible transparent conducting film

Dissolve BMI (0.05 g) firstly into 2 mL of DMF, and then add PU-Furan (0.60 g) into the solution. The mixture is magnetically stirred until a clear solution appears. Cast the solution onto a copper nanowire film coated glass slide. The liquid layer after evaporation of solvent is cross-linked at 80 °C for 2 days. Peel off the resulting Cu NWs/DA-PU composite film from the glass substrate (Figure 2.18).

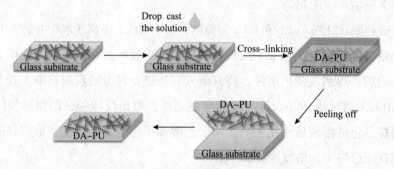

Figure 2.18 A schematic fabrication procedure of self-healing flexible transparent conductive composite film

4. Performance characterization and testing of self-healing flexible transparent conducting film

(1) Polarizing microscope observing

Observe the morphology of polyurethane film layed on glass slide by a polarizing microscope. Cut the film to a shallow crack, and then heat it at 120 ° C to observe the repair time and degree.

(2) Electrical properties testing of conductive films

The sheet resistance (Rsh) of composite conductive films can be measured using a 4-point probe sheet resistance meter. Firstly, the four pins were slowly went down to the surface of the composite conductive films. Then the voltage and current of the sample can be obtained from the sheet resistance meter. Calculate the sheet resistance of the composite conductive film from formula (1).

出载玻片，擦拭干净后放入丙酮中；3~5 min 后，用镊子取出，高速氮气吹干载玻片上的液体后备用。

②铜纳米线膜的制备：将制得的铜纳米线 CuNWs（4 mg/mL）超声分散后，用一次性针头注射器，取上述混合液 0.1 mL，将其全部滴在预先处理过的玻璃板上部，手握涂覆线棒，将铜纳米线均匀分散在玻璃板表面，盖上表面皿，待溶液中的溶剂正己烷挥发完毕。随后将铜纳米线膜置于管式炉中，在氢气氛围下，200 ℃退火处理，以减少接触电阻，增加电导率。

③自修复复合导电薄膜的制备：称取 0.05 g 4,4′-双马来酰亚胺基二苯甲烷（BMI）溶于 2 mLDMF，接着加入 0.6 g 以呋喃环为端基的聚氨酯。待两者充分混合后，将混合物均匀地铺在经退火处理之后的铜纳米线薄膜（含玻璃基质）上，在 80 ℃下反应 48 h，冷却至室温。此时的铜纳米线已经嵌在自修复聚氨酯内，将复合材料从玻璃基质上剥离即制得自修复复合导电材料（图 2.18）。

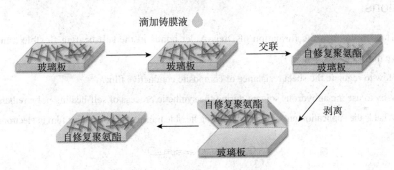

图 2.18　自修复复合导电材料制备步骤图

（4）自修复薄膜性能表征及检测。

①偏光显微镜测试。将铺在载玻片上的聚氨酯薄膜放置在显微镜的平台上进行观察，可观察制得的聚氨酯薄膜的形貌，用手术刀在薄膜上划一道浅浅的裂纹，在 120 ℃下加热观察修复时间及修复状况。

②导电薄膜的电学性能测试。导电复合材料的表面电阻测试主要是利用四探针测试仪，将四探针测试仪上的四个针头缓慢下降，直至接触含有铜纳米线膜的导电材料的表面，由与测试仪相连接的电压电流仪，分别读出样品的电压 V 和电流 I，通过式（1）可以计算得到导电材料的薄层电阻 R_{sh}。

$$R_{sh} = \frac{\pi}{\ln 2} \cdot \frac{V}{I} \tag{1}$$

Where, R_{sh} is sheet resistance ($\Omega \cdot sq^{-1}$), V is voltage and I is current.

(3) Transparency test of conductive film

Transmittance spectra of composite samples can be collected by using an INESA 721-100 visible spectrophotometer at room temperature across 350–850 nm using a 1 cm quartz cell.

(4) Simulation application of composite conductive film

A simple light-emitting diode (LED) circuit set is composed of one commercial LED light and two wires. A specimen is connected in series with a white light-emitting diode (LED) and a power source. The LED is initially lit as power was supplied to it. The resistance of the composite conductor is much smaller than that of the LED. After being cut across the width with a razor blade, the LED will be off because of the disconnection of power owing to the broken circuit. After that the specimen is then heated at 120 °C for 5 min, the LED is lit again with no noticeable difference in brightness. According to these phenomena the healing property of the composite conductor is presented.

Questions

(1) How to confirm the formation of Diels-Alder bonds in the self-healing flexible transparent conductive film?

(2) How to regulate the sheet resistance of composite conductive film?

(3) Why to use the anhydrous solvent during the synthetic process of self-healing polyurethane?

(4) What is the application prospect of self-healing flexible transparent conductive film in electronic field?

Experiment 18   Soap-Free Emulsion Polymerization of Styrene with the assistance of crown ethers

1. The purposes

(1) Learn the the method for performing soap-free emulsion polymerization and deepen the understanding of soap-free emulsion polymerization mechanism.

(2) Observe the influence of crown ethers on diameter of polystyrene nanoparticles, and learn how to characterize the nanoparticles.

(3) Explore the mechanisms of soap-free emulsion polymerization of styrene in the presence of crown ethers.

$$R_{sh}=\frac{\pi}{\ln 2}\cdot\frac{V}{I} \qquad (1)$$

式中，R_{sh} 为薄层电阻；V 为电压；I 为电流。

③导电薄膜透明度测试。自修复聚氨酯材料的透明度测试采用的是 721 型可见分光光度计，测试的波长范围为 350~850 nm，由仪器读出在一定波长下材料的透过率（T）。

④导电薄膜的模拟应用。

利用市售的 LED 灯和两根导线组成一组简单的 LED 电路：将样品薄膜与白色的 LED 灯及电源通过导线相互连接组成一通路。最开始时，LED 灯在电源的供应下保持亮着的状态。当导电复合薄膜被刀片划上一道与材料宽度相当的划痕时，由于回路断开，LED 灯也随之熄灭。将样品薄膜取下置于 120 ℃ 的条件下加热 5 min，LED 灯会再次亮起，由此可知柔性导电薄膜具有自修复性能。

思考题

1. 如何知道自修复导电薄膜中 Diels–Alder 键的存在？
2. 自修复导电薄膜的导电性能如何调控？
3. 自修复聚氨酯的合成过程中使用无水溶剂的原因？
4. 柔性透明自修复薄膜在电子领域的应用前景如何？

实验十八　冠醚辅助苯乙烯的无皂乳液聚合

一、实验目的

（1）通过实验，学习无皂乳液聚合的方法，加深对无皂乳液聚合的理解。

（2）观察冠醚的加入对聚苯乙烯纳米球粒径的影响，学会粒径表征方法。

（3）探讨冠醚存在下聚苯乙烯纳米球的成核机理。

2. Experimental principle

It is necessary to add large amount of surfactane for the conventional emulsion polymerization process, which will deteriorate the electrical, optical, surface and film-forming properties of the obtained particles. Also, the surfactants usually cause serious environmental pollution. In contrary, soap-free emulsion polymerization technique can prepare clean and narrow dispersed monodisperse nanoparticles without needing large amounts of surfactants. There are two kinds of acceptable nucleation mechanisms of soap free emulsion polymerization: homogeneous nucleation mechanism and oligomer micelle nucleation mechanism. Commonly, homogeneous nucleation and oligomer micelle nucleation mechanisms are suitable for hydrophilic monomer and hydrophobic monomers, respectively. Styrene is a hydrophobic monomer, and it will first produce surface active oligomer chains terminated with initiator anions in aqueous phase. When the concentration of oligomers reaches the critical value, the oligomer micelles are formed. The formed oligomer micelles have solubilization to styrene and further initiate the monomer reaction to form primary particles. The unstable primary particles in the water phase will aggregate into the final nanoparticles.

Crown ethers are one kind of cyclic oligomers of ethylene oxide with hydrophobic exterior rings and hydrophilic interior rings. They are usually defined as phase transfer catalyst (PTC) because the hydrophilic interior rings can strongly bind certain cations to form complexes, e.g. 18-crown-6 for potassium. In this experiment, two significantly different sizes of polystyrene nanoparticles were synthesized with the addition of 18-crown-6 and 12-crown-4 in the soap-free emulsion polymerization of styrene. There are two roles for crown ether in the soap-free emulsion polymerization of styrene. On one hand, crown ethers as phase catalyst could facilitate the polymerization rate of styrene. In addition, the hydrophobic exterior ring of crown ethers probably increase the solubility of styrene and the critical chain length of PS in aqueous phase based on "like dissolve like" rule. On the other hand, the presence of crown ethers in monomer/water interface may form a physical barrier and restrict the transfer of monomer from monomer phase to the PS oligomers, which can inhibit the rate of polymerization and hinder the growth and coagulation of polymer chains. The former will increase the size of PS nanoparticles, while the latter will inhibit the size growth of PS nanoparticles.

3. Instruments and reagents

Instruments: Gas chromatography; transmission electron microscope (TEM); fourier transform infrared spectrum (FT-IR); dynamic light scattering (DLS); water bath; mechanical stirrer; desktop centrifuge; electronic balance; vaccum drying oven; ultrasonic bath; three –neck round bottom flask (250 mL); beaker (100 mL).

Reagents: styrene; sodium hydroxide; 18-crown-6; 12-crown-4; potassium persulfate; nitrogen; hexanol; deionized water; dichloromethane.

二、实验原理

传统的乳液聚合需要添加大量乳化剂，反应后会残留在纳米粒子表面，对聚合产物的电性能、光学性能、表面性能及成膜性能等带来不良影响，并且乳化剂还会造成环境污染。无皂乳液聚合方法不需要添加大量乳化剂，能够制备出表面干净、粒径分布窄的单分散纳米粒子，应用前景广阔。关于无皂乳液聚合的成核机理一般认为有两种：均相凝聚成核机理和齐聚物胶束成核机理。均相成核机理适用于水溶性较大的单体，而齐聚物胶束成核机理则适合水溶性小或疏水性单体。苯乙烯属于疏水性单体，它在无皂乳液聚合反应中首先会在水相中产生一端带有引发剂阴离子因而具有表面活性的低聚物链，低聚物链浓度达到临界值时形成低聚物胶束，低聚物胶束对单体具有增溶作用进一步引发单体反应形成初级粒子，初级粒子在水相中不稳定进一步聚集成最终的纳米粒子。

冠醚是具有疏水性外环和亲水性内环的环氧乙烷的环状低聚物，经常作为相转移催化剂（PTC）使用，因为亲水性内环可以牢固地结合某些阳离子以形成络合物。本实验通过在苯乙烯的无皂乳液聚合体系中添加18冠醚和12冠醚，得到了粒径截然不同的PS纳米颗粒。冠醚在苯乙烯的无皂乳液聚合体系中起到了两方面作用，一是作为相转移催化剂促进了聚苯乙烯高分子链的增长，并且冠醚的疏水外环增加了PS的临界链长；另一方面冠醚的存在可能起到了阻碍单体向乳胶核扩散的物理屏障作用。前者占优势会得到增加PS纳米粒子的尺寸，后者占优势则会抑制PS纳米粒子尺寸的增长。

三、仪器和试剂

仪器材料：气相色谱仪；透射电镜显微镜；傅里叶红外光谱；动态光散射仪；顶置式电子搅拌器；水浴锅；离心机；电子天平；真空干燥箱；超声波清洗器；三口烧瓶（250 mL）；烧杯（100 mL）。

试剂：苯乙烯；氢氧化钠；十八冠醚；十二冠醚；过硫酸钾；氮气；去离子水；正己醇；二氯甲烷。

4. Experimental procedure

(1) Purification of styrene

The inhibitor, hydroquinone, should be removed from styrene by basic aluminum oxide flash column before performing soap-free emulsion polymerization. Briefly, a glass burette with absorbent cotton on the tip is packed with basic aluminum oxide. The height of the self-made column is about 10 cm. 3 mL of styrene was passed through the column by using a rubber suction bulb. Collect the purified styrene for further polymerization.

(2) Weighing

68 mg of potassium persulfate, 50 mL of deionized water and certain amounts of crown ether are added to a 100 mL beaker to form aqueous phase. Then, the aqueous phase is transferred to a three-neck flask fitted with a reflux condenser, a mechanical stirrer and a nitrogen inlet nozzle. Finally, 2 mL of purified styrene is added to the flask with pipette. The amount of 18-crown-6 added is 0.33 g, and 12-crown-4 is 0.213 g. there is no crown ether in the control experiment.

(3) Polymerization

The mixture is stirred at 200 rpm to form a heterogeneous system and then bubbled with nitrogen to remove oxygen for 30 min. Finally, the temperature is elevated to 75 °C for 6 h of polymerization. Schematic illustration of experimental device is shown in Figure 2.19.

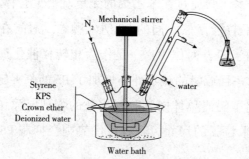

Figure 2.19　Schematic illustration of experimental device of styrene soap-free polymerization

(4) Sample processing

The PS nanoparticles as prepared are obtained by centrifuging for 15 min, and then the nanoparticles are washed by water and ethanol for two times in turn and dried in a vacuum drying oven at room temperature.

(5) Characterization

The morphology and size of polystyrene nanoparticles are characterized by a transmission electron

四、实验步骤

（1）苯乙烯的精制。

在进行无皂乳液聚合前，采用碱性三氧化二铝柱去除单体苯乙烯中的阻聚剂对苯二酚，具体操作如下：取一根玻璃滴定管，在尖端塞上脱脂棉作为筛板，然后装入碱性三氧化二铝，高度 10 cm 左右。用吸管从试剂瓶中取 3 mL 的苯乙烯加入自制的玻璃柱中，用吸耳球轻轻挤压，收集脱除阻聚剂的苯乙烯后备用。

（2）称量。

将 68 mg 过硫酸钾，50 mL 去离子水和一定量的冠醚置于 100 mL 小烧杯中，溶解后加入装有回流冷凝器和机械搅拌器的三口烧瓶中，用移液枪加入 2 mL 精制好的苯乙烯。其中空白对照实验不加冠醚，18 冠醚的加入量为 0.33 g，12 冠醚的加入量为 0.213 g。

（3）聚合反应。

控制搅拌速度 200 rpm，通氮气鼓泡以除氧 30min。最后，将反应体系温度升高至 75 ℃聚合 6 h。实验装置图如图 2.19 所示。

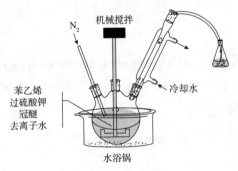

图 2.19　实验装置图

（4）样品处理。

反应结束后采用高速离心机离心 15min 获得聚苯乙烯纳米颗粒，然后依次用水和乙醇洗涤纳米颗粒两次，并在室温下真空干燥。

（5）样品表征。

形貌观察：将纳米颗粒乙醇分散液滴到镀有碳膜的铜网上，并在室温下

microscopy (TEM) operating at 120 KV. One drop of the nanoparticles/ethanol mixture is deposited on the carbon membrane of the copper grid, and the ethanol is evaporated at room temperature. The average diameter of the nanoparticles is analyzed from TEM image of about 50 particles using Nanomeasure software. The coefficient of variation of particle size distribution (CV) is defined by the following equation (1):

$$CV = \left(\sum_{i=1}^{n} \frac{(d_i - \bar{d})^2}{N} \right)^{\frac{1}{2}} / \bar{d} \qquad (1)$$

Where, N is the total number of particles, d_i is the diameter of single particle, $\bar{d}$ is the average diameter of particles.

Tetrahedral transparent quartz cuvette is used for the measurement of hydrodynamic radius of polystyrene nanoparticles on a dynamic light scattering.

Monomer conversion is measured by a gas chromatography and the injection temperature is 300 °C. Take 1 mL of reaction mixture from the three-neck flask at the prescribed time and dilute it 4 times with deionized water, and then the mixture is extracted by 2 mL of dichloromethane to obtain the residual styrene. 1 μL of the extract is injected and hexanol is used as an internal standard (0.1%). The monomer conversion is obtained by calculating the peak area of styrene before and after polymerization at specified time.

In order to observe whether or not the incorporation of crown ethers in the surface of polystyrene nanoparticles, the composition of polystyrene nanoparticles with/without crown ethers can be analyzed by Fourier transform infrared spectroscopy using KBr pellet method. The spectrum is recorded at 400-4000 cm^{-1}.

Questions

1. Why the addition of crown ethers has significant effect on the size of polystyrene nanoparticles?

2. What are the effects of 18-crown-6 and 12-crown-4 on the nucleation mechanism of styrene soap-free emulsion polymerization?

Experiment 19 Synthesis of polyacrylamide with ultrahigh molecular weight

1. The purpose

(1) Understand the method and principle of aqueous solution polymerization.

(2) Master the operation of anaerobic reaction.

蒸发乙醇，采用透射电镜（TEM）观察 PS 纳米粒子形貌和尺寸，操作电压 120 KV。使用 Nanomeasure 软件从约 50 个颗粒的 TEM 图像分析了纳米颗粒的平均粒径。粒度分布的变异系数（CV）由式（1）可得：

$$CV=\left(\sum_{i=1}^n \frac{(d_i-\bar{d})^2}{N}\right)^{\frac{1}{2}}/\bar{d} \qquad (1)$$

式中，N 是颗粒总数；d_i 是单个颗粒的直径；$\bar{d}$ 是平均粒径。

聚苯乙烯纳米粒子的尺寸还可以通过动态光散射仪来测量：取适量样品用去离子水稀释，加入四面透光的比色皿中，用动态光散射仪得到纳米粒子的水力学半径。

单体转化率：苯乙烯单体转化率通过气相色谱仪进行测量，进样温度 300℃。样品制备方法：在规定的时间从三口烧瓶中取出 1 mL 反应混合物，并用去离子水稀释 4 倍，然后将混合物用 2 mL 二氯甲烷萃取，得到残留的苯乙烯。注入 1μL 提取物，并用正己醇用作内标物（0.1 %）。通过在特定时间聚合前后聚苯乙烯的峰面积来获得单体转化率。

采用 KBr 压片法，通过傅里叶变换红外光谱法分析聚苯乙烯纳米粒子的化学组成，记录 400~4000 cm^{-1} 处光谱数据。观察聚苯乙烯纳米粒子表面是否存在冠醚。

思考题

1. 冠醚的加入为什么会对聚合物粒径产生很大的影响？
2. 两种冠醚加入后对苯乙烯无皂乳液聚合成核机理的影响。

实验十九　超高分子量聚丙烯酰胺的合成

一、实验目的

1. 了解水溶液聚合的方法和原理。
2. 掌握除氧反应的操作。

(3) Synthesize polyacrylamide with ultra-high molecular weight.

(4) Learn how to use combinational initiators and understand their respective roles.

2. Experimental principle

Solution polymerization refers to the polymerization that occurred in a monomer dissolved solvent. Some of the resulting polymers are soluble in solvent, and somw are insoluble. The former is called homogeneous polymerization, and the latter is called precipitation polymerization. This experiment and experiment 7 (acrylamide aqueous solution polymerization) both belong to homogeneous polymerization, but the difference between the two experiments lies in differnet initiating system.

In Experiment 7, ammonium persulfate is the only nitiator. Because water is good solvent for polyacrylamide, the polymer chain is in a relatively stretched state and the degree of chain wrapping is relative shallow. The chain segment is easy to diffuse and result in diradical termination of the active end group of polymer chains. According to the diradical termination mechanism, the polymerization rate is proportional to the square root of the initiator concentration. Because the monomer concentration is not high enough, it may eliminate the auto-accelerating effect and make the reaction follow the normal free radical polymerization kinetics. Therefore, the experiment belongs to constant temperature reaction at a certain temperature.

The initiators used in this experiment is a composite initiator system, and each initiator functions at different temperature during the polymerization process. The starting polymerization temperature is 0-10 ℃, a very small amount of redox initiator initiates polymerization at this stage. The exothermic heat of polymerization will increase the system temperature to about 30 ℃, and the water-soluble azo initiator starts to work. When the temperature increases to 50 ℃, the peroxide initiator starts to work, and the polymerization continues to the end. Since the concentration of initiator in each stage is low, which favors to synthesis of high molecular weight polymer, and finally higher temperature helps to the complete conversion of monomer. Since the initiation dose in each polymerization stage is less, and the concentration of monomer is relatively high, the phenomenon of automatic acceleration appears. In this experiment ultrahigh molecular weight polyacrylamide can be obtained without heating, but oxygen has inhibitory effect on this experiment. Therefore, nitrogen gas should be introduced to eliminate the inhibitory effect of oxygen.

In this experiment, the self-accelerating phenomenon can be observed by recording the T-t relationship. The heat of polymerization of the reaction can be calculated according to the T_0-T_e. Since acrylamide is a water-soluble monomer, the solution reaction using water as a solvent has the advantage of being inexpensive and non-toxic.

3. Instruments and reagents

Instruments: Electronic balance; vacuum drying oven; timer; mineral water bottle with three holes in the bottle cap; thermometer; long glass tube; high-purity nitrogen; latex pipe.

3. 合成一种具有超高分子量的聚丙烯酰胺。
4. 学习组合引发剂的作用及使用方法。

二、实验原理

单体溶于溶剂中而进行的聚合方法叫作溶液聚合。生成的聚合物有的在溶剂中溶解，有的在溶剂中沉淀。前一种情况称为均相聚合，后一种叫作沉淀聚合。本实验与实验七（丙烯酰胺水溶液聚合）都属于均相聚合，但两次实验的不同之处在于引发剂的不同。

在实验七中，引发剂仅有过硫酸铵。由于水是聚丙烯酰胺的良溶剂，聚合物链处于比较伸展的状态，包裹程度较浅，链段扩散容易，活性端基容易相互靠近而发生自由基双基终止反应。依双基终止机理，聚合速率与引发剂浓度的平方根成正比，又由于单体的浓度不够高，则可能消除自动加速效应，使反应遵循正常的自由基聚合动力学规律。因此实验属于在一定温度下的恒温反应。

本实验中所用的引发剂是复合引发剂，各引发剂在不同的聚合温度阶段起作用：反应起始温度 0~10℃，极少量的氧化还原引发剂引发聚合，聚合放热将反应温度升高到 30 ℃左右，水溶性偶氮引发剂开始起作用，继续聚合将温度升到 50℃，过氧化物引发剂开始起作用，继续聚合至结束。在每一段起作用的引发剂浓度都较低，有利于生成高分子量聚合物，最后较高温度有利于单体的完全转化。由于在每个聚合阶段的引发剂量都较少，而单体浓度相对较高，因此出现自动加速现象。本实验不需加热即可生成超高分子量聚合物，但氧气对本实验也具有阻聚作用，因此需通入氮气，从而消除氧气的阻聚作用。

本实验可通过记录反应温度－时间（T-t）关系，观察实验自加速现象。并可根据 T_0-$T_终$，计算反应的聚合热，由于丙烯酰胺为水溶性单体，因此采用水为溶剂进行溶液反应，其优点是价廉无毒。

三、仪器和试剂

仪器：电子天平；真空干燥箱；计时器；瓶盖上带孔的矿泉水瓶；温度计；长玻璃管；乳胶管。

Reagents: Acrylamide; potassium persulfate (KPS); 2,2'-azobis[2-(2-imidazolin-2-yl) propane] dihydrochloride (VA-044); sodium bisulfite ($NaHSO_3$); deionized water; high purity nitrogen.

4. Experimental procedure

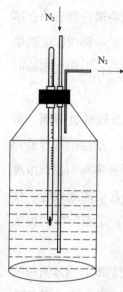

Figure 2.20 Schematic of polymerization device

(1) Prick 3 holes on the lid of the mineral water bottle, including 2 large holes and 1 small hole. Insert a glass tube and a thermometer with a rubber sleeve respectively into these two large holes, and promise to plug the two holes tightly. Figure 2.20 shows the schematic of polymerization device❶.

(2) Add 50 g of acrylamide and 150 g of water to the bottle. The glass tube should be extended to the bottom of the bottle and the thermometer shall be extended below the liquid level. Connect the glass tube to the double row tube and put it into high-purity nitrogen for 20 minutes❷, and cool the reaction bottle to about 5 ℃ with ice water bath.

(3) Add 3 mL of 1% (w/w) peroxide initiator $K_2S_2O_8$ aqueous solution to the reaction bottle, continue to inject nitrogen for 5 minutes, and then add 1.5 mL of 1% (w/w) azo initiator VA-044 aqueous solution to the reaction bottle. Continue to bubble nitrogen for 10 minutes, and finally add 0.3 mL of 1% (w/w) $NaHSO_3$ aqueous solution❸.

(4) Remove the reaction bottle from the ice water bath. The reaction bottle is coated with foam plastic film and insulated from the outside. Record the temperature changes with reaction time. Observe whether the liquid in the reaction bottle becomes viscous, that is, whether it polymerizes. In case of polymerization, the viscosity of the reaction system will increase, and the bubble of nitrogen will rise slowly, so nitrogen can be stopped.

(5) When the reaction temperature stops rising and starts to fall, the reaction ends. Cut open the mineral water bottle to get the rubber block, which can calculate the solid content, monomer conversion and molecular weight of polyacrylamide after vacuum drying.

Attentions: In the above steps 2 and 3, keep nitrogen flowing continuously. When adding the initiator, open the bottle cap and add it quickly, then cover it quickly.

Questions

(1) Try to calculate the heat of polymerization per mole of molecule.

---

❶ In this experiment, a combinational initiator system is used to synthesize ultrahigh molecular weight polymers.

❷ This experiment considers the inhibitory effect of oxygen, and reduces or eliminates the inhibitory effect of oxygen by bubbling nitrogen.

❸ The glass tube should go deep into the bottom of the reaction bottle to replace the upper air.

试剂材料：丙烯酰胺；过硫酸钾（$K_2S_2O_8$）；偶氮二异丁咪唑啉盐酸盐（VA-044）；亚硫酸氢钠（$NaHSO_3$）；去离子水；高纯氮气。

四、实验步骤

（1）在矿泉水瓶盖上扎3个孔，其中2个大孔，1个小孔。将带有橡皮套的玻璃管和温度计插入这2个大孔中，将其堵严，实验装置如图2.20所示❶。

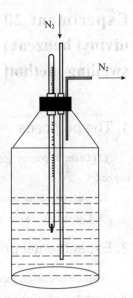

（2）向瓶中加入50 g丙烯酰胺和150 g水，并使玻璃管伸入瓶底，温度计伸入液面以下，将玻璃管连接双排管通入高纯氮气20min，并用冰水浴使反应瓶降温至5 ℃左右❷。

（3）向反应瓶中加入1%（质量）过氧化物引发剂$K_2S_2O_8$水溶液3 mL，继续通入氮气5min；再向反应瓶中加入1%（质量）偶氮引发剂VA-044水溶液1.5 mL。继续通氮气10min；最后加入1%（质量）还原剂$NaHSO_3$水溶液0.3 mL❸。

图2.20 聚合装置图

（4）将反应瓶从冰水浴中取出，反应瓶用泡沫塑料膜包覆与外界绝热。记录反应时间与温度变化。观察反应瓶中液体是否变黏稠，即是否聚合。若聚合则反应体系黏度增加，氮气气泡上升缓慢，可停止通氮气。

（5）当反应温度停止上升并且开始下降，反应结束。剪开矿泉水瓶，得到胶块，真空干燥后可以测试固体含量、转化率以及分子量。

注意：步骤（2）和步骤（3）中要保持一直通氮气。加入引发剂时，要打开瓶盖后迅速加入，然后迅速盖上。

思考题

1. 尝试计算每摩尔分子的聚合热。

---

❶ 本实验采用加入组合引发剂，从而合成超高分子量的聚合物。
❷ 本实验考虑氧气的阻聚作用，通过通氮气减少或消除氧气阻聚作用。
❸ 玻璃管中应深入反应瓶底部，从而排出上层的空气。

(2) Why there is a self-accelerating effect of the reaction?

Experiment 20 Synthesis of gigaporous poly(styrene-divinyl benzene) microspheres by surfactant reverse micelles swelling method

1. The purposes

(1) Learn to prepare gigaporous polystyrene microspheres by surfactant reverse micelles swelling method.

(2) Understand the mechanism of gigapore formed by swelling of surfactant reverse micelles.

(3) Understand the characterization methods of microsphere morphology, particle size and pore size

2. Experimental principle

Suspension polymerization refers to initiator contained monomer is suspended in water in the form of droplets for radical polymerization. Overall, this system is oil in water (O/W) emulsion, in which water is the continuous phase and monomer is the dispersed phase. Because the polymerization takes place in each droplet, the reaction mechanism is the same as bulk polymerization, which can be regarded as bulk polymerization of beads. It can also be divided into homogeneous and heterogeneous polymerization according to the solubility of the polymer in the monomer. If a water-soluble monomer containing initiator is suspended an oil continuous phase to carry out polymerization, this method is called inverse suspension polymerization.

The preparation of gigaporous polymer microspheres by reverse micelle swelling method is the same as that of conventional suspension polymerization, and the preparation process is very simple. The difference is that a higher content of surfactant is added to the oil phase. When the surfactant concentration in the oil phase increases to the critical micelle concentration (CMC), the surfactant molecules will self-assembly into the micelle to absorb water and swell. The schematic diagram of gigaporous microspheres prepared by the reverse micelle swelling method is shown in Figure 2.21.

3. Instruments and reagents

Instruments: Optical microscopy, scanning electron microscopy (SEM); laser particle size analyzer, mercury porosimetry, circulating water vacuum pump, vaccum drying oven, electronic balance, four-neck round bottom flask (250 mL), thermometer; mechanical stirrer, soxhlet extractor, water bath, triangular flask (250 mL), beaker (100 mL), G3 sinter funnel, glass slide.

2. 为什么会出现反应的自加速效应?

实验二十　表面活性剂反胶团溶胀法制备超大孔聚苯乙烯微球

一、实验目的

（1）学习利用表面活性剂反胶团溶胀法制备超大孔聚苯乙烯微球。
（2）理解表面活性剂反胶团溶胀形成超大孔的机理。
（3）了解微球形貌、粒径及孔径的表征手段。

二、实验原理

含有引发剂的单体以液滴状悬浮于水中进行自由基聚合的方法称为悬浮聚合法。整体看该体系是水包油（O/W）乳液，水为连续相，单体为分散相。聚合在每个小液滴内进行，反应机理与本体聚合相同，可看作小珠本体聚合。同样也可根据聚合物在单体中的溶解性有均相、非均相聚合之分。如果将水溶性单体的水溶液作为分散相悬浮于油类连续相中，在引发剂的作用下进行聚合的方法，称为反相悬浮聚合法。

利用表面活性剂反胶团溶胀法制备超大孔聚合物微球，这种方法的操作步骤与常规的悬浮聚合相同，制备过程非常简单，不同点是在油相中添加了较高含量的表面活性剂。当油相中的表面活性剂浓度增加到临界胶束浓度（CMC）时，表面活性剂分子会聚集成胶束进而吸水溶胀。反胶团溶胀法制备超大孔微球的示意图如图 2.21 所示。

三、仪器和试剂

仪器：光学显微镜；扫描电镜；激光粒度仪；压汞仪；循环水真空泵；真空干燥箱；电子天平；四口烧瓶（250 mL）；温度计；顶置式电子搅拌器；索氏抽提器；水浴锅；电子天平；烧杯（100 mL）；锥形瓶（500 mL）；G3 砂芯漏斗；载玻片。

Polymer Chemistry Experiments (Bilingual)
高分子化学实验

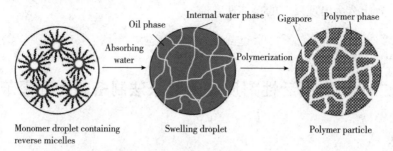

Figure 2.21　Schematic diagram of gigaporous microspheres prepared by the reverse micelle swelling method

Reagents: styrene; divinyl benzene; benzoyl peroxide (BPO); hexadecane (HD); sorbitan monooleate (Span 80); poly (vinyl alcohol) (PVA217); hydroquinone (HQ); sodium dodecyl sulfate (SDS); sodium sulfate; ethanol; deionized water; high purifity nitrogen.

4. Experimental procedure

(1) Remove inhibitor hydroquinone from styrene and divinylbenzene by alkaline alumina column[❶]. Briefly, a glass burette with absorbent cotton on the tip is packed with basic aluminum oxide. The height of the self-made column was about 12 cm. 5 mL of styrene is passed through the column by using a rubber suction bulb. The purified styrene is collected for next polymerization. The refining method of divinylbenzene is the same as that of styrene.

(2) Because PVA is insoluble in cold water, it should be dissolved in advance and then diluted for use. Add 10 g PVA and 90 mL of deionized water to a conical flask, then gradually raise the temperature to the boiling state under the stirring state till PVA is completely dissolved, and cool it for backup.

(3) Take three beakers (100 mL) for preparation of oil phase with different amounts of surfactant. The oil phase is composed of monomer (st), crosslinker (DVB), diluent (HD), surfactant (Span 80) and initiator (BPO). Weigh each component accurately by electronic balance, put them into a beaker and stir to dissolve. The water phase is composed of stabilizer (PVA), surfactant (SDS), Na_2SO_4, inhibitor (HQ) and deionized water. The water phase components are directly added into the four-neck round bottom flask (Figure 2.22). See Table 2.7 for the specific composition of oil phase and water phase. Finally, the oil phase is dispersed into water phase to make O/W emulsion. The stirring speed is 160 r/min. Pick a small amount of emulsion from the flask and put it on a glass slide to observe the emulsion state under the optical

---

[❶]　Styrene (St) and crosslinker divinylbenzene (DVB) are industrial pure. The purpose of refining before use is to remove the polymerization inhibitor contained in the reagents.

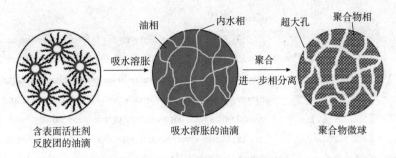

图 2.21 聚合物微球超大孔形成机理示意图

试剂材料：苯乙烯（St）；二乙烯基（DVB）；过氧化苯甲酰（BPO）；十六烷（HD）；Span80；聚乙烯醇（PVA）；对苯二酚（HQ）；十二烷基硫酸钠（SDS）；硫酸钠（Na_2SO_4）；乙醇；去离子水；高纯氮气。

四、实验步骤

（1）采用碱性三氧化二铝柱去除单体苯乙烯和二乙烯基苯中的阻聚剂对苯二酚❶。具体操作如下：取一根玻璃滴定管，在尖端塞上脱脂棉作为筛板，然后装入碱性三氧化二铝，高度 12 cm 左右。用吸管从试剂瓶中取 5 mL 的苯乙烯加入自制的玻璃柱中，用吸耳球轻轻挤压，收集脱除阻聚剂的苯乙烯后备用。二乙烯基苯的精制方法同苯乙烯。

（2）由于 PVA 不溶于冷水，需要提前将其溶解后稀释使用。量取 90 mL 去离子水置于锥形瓶中，在搅拌状态下加入 10 g PVA，逐渐升温至接近沸腾状态使 PVA 完全溶解，冷却后备用。

（3）准备三个烧杯，洗净、烘干，用来配制不同表面活性剂含量的油相。油相是由单体（St）、交联剂（DVB）、稀释剂（HD）、表面活性剂（Span 80）和引发剂（BPO）组成，使用电子天平精准的称取各组分，加入到烧杯中完全溶解。水相是由稳定剂（PVA）、表面活性剂（SDS）、Na_2SO_4、抑制剂（HQ）和去离子水组成，水相组分直接加入四口烧瓶中，实验装置如图 2.22 所示。油相和水相的具体组成见表 2.7，把油相分散于水相制成 O/W 乳液，搅拌桨转速设置为 160 r/min，用吸管吸取少量乳液置于载玻片上，置于光学显微镜

---

❶ 苯乙烯（St）和交联剂二乙烯基苯（DVB）为工业纯，在使用前精制是为了除试剂中含有的阻聚剂。

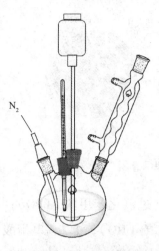

Figure 2.22 Schematic diagram of experimental device

microscope ❶.

(4) After bubbling nitrogen for 0.5 h ❷, raise the temperature of water bath to 75 ℃ and start the polymerization. After 20 hours polymerization under nitrogen atmosphere, gigaporous polystyrene microspheres were filtered with G3 sinter funnel, and washed several times with hot water and ethanol respectively.

(5) Remove unreacted monomer (St and DVB), surfactant (Span 80) and solvent (HD) in the particles using ethanol as extraction solvent by Soxhlet extractor. The extraction time is 6 h.

(6) The final product is obtained after vacuum drying at room temperature. The product yield can be calculated from the mass of the dried microspheres, as shown in the following formula (1).

$$y = \frac{m_p}{m} \times 100\% \tag{1}$$

Where, m_p is the mass of the microspheres after drying, and m is the total mass of St and DVB.

(7) The surface morphology of the microspheres can be observed using scanning electron microscopy (SEM). First, disperse the microspheres in ethanol (concentration is about 5 mg/mL), and then drop them onto aluminum foil. After drying at ambient atmosphere, take a small piece of aluminum foil and fix it on the sample stub with conductive double-sided tape. After the sample stub is sprayed with gold on the vacuum ion sputtering machine, put it into the electron microscope sample chamber for observation.

(8) The particle size distribution of the polystyrene microspheres is measured by a laser particle size analyzer, that is, a certain amount of microspheres are suspended in deionized water with a concentration in the range of 1-2 mg/mL. After ultrasonic dispersion, it is added to a laser particle size analyzer to determine the particle size distribution.

(9) The structure of the microspheres such as porosity, pore size distribution, specific surface area and other data are determined by the mercury porosimeter.

Table 2.7 Standard recipe for preparation of gigaporous polystyrene microspheres

Ingredients	Weight /g
Continuous phase	
PVA	1.0
HQ	0.01

---

❶ The particle size can be predicted by size of oil droplets in emulsion using optical microscope, and the particle size of microspheres can be adjusted by adjusting the stirring speed.

❷ The inhibition of O_2 to polymerization can be reduced or eliminated by bubbling N_2. N_2 glass tube shall extend into the bottom of the reaction bottle to discharge the air in the solution.

下观察乳液状态❶。

（4）通氮气 0.5 h 后❷，将温度升至 75 ℃开始聚合。氮气气氛下聚合 20 h，用 G3 砂芯漏斗过滤得到聚合物微球，并分别用热水和乙醇清洗数次。

（5）用索氏抽提器除去微球内未反应的单体（苯乙烯和二乙烯基苯）以及未参与聚合的物质如表面活性剂（Span 80）、溶剂（十六烷）等，抽提溶剂为乙醇，抽提时间 6 h。

（6）室温真空干燥后，即得成品。产品收率由干燥后微球的质量计算而得，如式（1）所示。

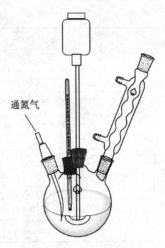

图 2.22　实验装置图

$$y = \frac{m_p}{m} \times 100\% \qquad (1)$$

式中，m_p 为干燥后微球的质量，m 为 St 和 DVB 的总质量。

（7）采用扫描电子显微镜对微球的表面形态进行观察。首先，将微球分散在乙醇中（浓度约为 5 mg/mL），再将其滴加到铝箔纸上，晾干后，取一小块，用导电双面胶将其固定在样品台上，将样品台在真空离子溅射机上喷金后，放入电镜样品室中进行观察。

（8）微球的粒径分布采用激光粒度仪测定，即将一定量的微球悬浮于去离子水中，浓度在 1~2 mg/mL 范围内。超声分散后加到激光粒度仪中进行粒径分布测定。

（9）微球的结构如孔隙率、孔径分布、比表面积等数据由压汞仪测定。

表 2.7　超大孔聚苯乙烯微球制备标准配方

配料	质量 /g
水相	
PVA	1.0
HQ	0.01

---

❶ 通过显微镜观察乳液中油滴的大小可以预测微球的粒径，通过调控搅拌速度调节微球的粒径大小。

❷ 考虑到 O_2 的阻聚作用，通 N_2 可以减少或消除 O_2 阻聚作用。N_2 玻璃管应伸入反应瓶底部，从而排出溶液中的空气。

续表

Ingredients	Weight /g
Na$_2$SO$_4$	0.02
SDS	0.015
Water	100
Dispersion phase	
BPO	0.16
Total monomer (St and DVB)	4.0
DVB	1.0
HD	0.2
Span 80	1.2, 1.6, 2.0

Questions

(1) What is the physical meaning of HLB value of surfactant?

(2) What is the effect of surfactant Span80 content in oil phase on the morphology and yield of polymer microspheres?

(3) In the experiment, the HLB value of Span80 is 4.3. What is the effect of surfactant HLB value on the morphology and pore size of microspheres?

Experiment 21   Synthesis of thermosensitive high-temperature resistant superabsorbent via solution polymerization

1.The purpose

(1) Learn to synthesize thermosensitive high-temperature resistant superabsorbent via solution polymerization.

(2) Understand the swelling mechanism of thermosensitive high-temperature resistant superabsorbent at high temperature.

(3) Master the property and structure characterization methods of thermosensitive high-temperature resistant superabsorbent.

2.Experimental principle

Superabsorbents, also known as absorbent resins, are very frequently used functional polymer materials. Compared with their own mass, they can absorb large amounts of water in aqueous solution.

续表

配料	质量/g
Na_2SO_4	0.02
SDS	0.015
水	100
油相	
BPO	0.16
St	3.0
DVB	1.0
HD	0.2
Span 80	1.2, 1.6, 2.0

思考题

（1）什么是 HLB 值？表面活性剂 HLB 值的物理意义。

（2）通过实验观察，可以得出油相中表面活性剂 Span 80 含量对聚合物微球的形貌、收率有什么影响？

（3）实验中 Span80 的 HLB 值为 4.3，表面活性剂的 HLB 值对微球的形貌和孔径有什么影响？

实验二十一　溶液聚合法制备温敏性耐高温体膨颗粒

一、实验目的

（1）学习利用溶液聚合制备温敏性耐高温体膨颗粒的方法。
（2）理解温敏性耐高温体膨颗粒高温下吸水膨胀机理。
（3）掌握温敏性耐高温体膨颗粒的性能及结构表征手段。

二、实验原理

体膨颗粒，又叫吸水性树脂，是使用非常频繁的功能性高分子材料，相比它自身重量而言，其在水溶液中会吸收大量水分。拥有超高吸水性能的体膨颗粒在去离子水中的膨胀倍率能达到 10~1000 倍，然而普通颗粒的吸水能

The swelling ratios of superabsorbents with super high water absorption can reach 10-1000 times in deionized water, whereas the water absorption capacity of ordinary particles is not more than 1 times. According to the electric charge property of polymer chain, superabsorbents can be divided into nonionic, ionic (including anionic and cationic) and zwitterionic substances. Most of the commercial superabsorbents are anionic. On the basis of monomer structure, the superabsorbents can be divided into crosslinked polyacrylic acid and polyacrylamide series, hydrolytic swelling polyacrylonitrile series and crosslinked maleic anhydride series. In addition to artificial synthesis, there are natural superabsorbents based on polysaccharides and peptides. It should be noted that in most cases, superabsorbents refer to partially neutralized acrylic acid (AA) or acrylamide (AM). Superabsorbents are widely used in the fields of health, agriculture and forestry, oil field and construction.

Based on the understanding of high temperature cleavage and hydrolysis of amide bond, thermosensitive superabsorbents are prepared with complex crosslinkers: high temperature resistant crosslinker (tetraallylammonium chloride) and N,N-methylene bisacrylamide that nontolerant to high temperature. The swelling mechanism of the thermosensitive high-temperature superabsorbent is shown in Figure 2.23. At low temperature, the two crosslinkers of superabsorbent act at the same time, resulting in high crosslinking density and low swelling ratio of the superabsorbent. Under high temperature, the crosslinking amide bond formed by N,N-methylene bisacrylamide is hydrolyzed. The crosslinking density decreases and the swelling ratio of the superabsorbent will greatly improve.

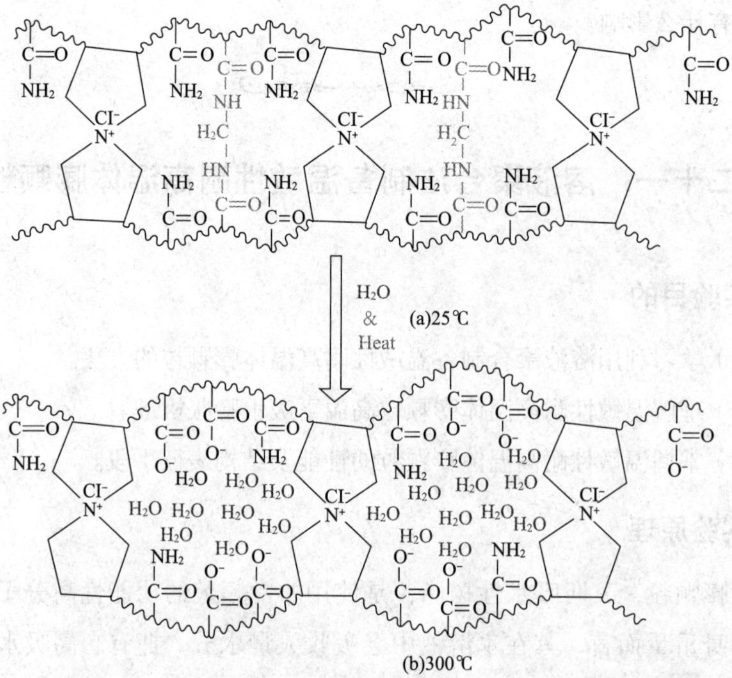

Figure 2.23　Schematic of swelling mechanism of thermosensitive high-temperature resistant superabsorbent

力却不会超过 1 倍。按照高分子链上电荷电性可将体膨颗粒分为非离子型、离子型（包括阴离子类型和阳离子类型）以及两性离子型的物质，大多数商用体膨颗粒都是阴离子型的。根据组成其化学结构的单体分类，体膨颗粒可分为交联的聚丙烯酸和聚丙烯酰胺系列、依靠水解膨胀的聚丙烯腈系列以及交联的马来酸酐系列。除了人工合成，还有一类天然的体膨颗粒，即基于多糖和多肽的体膨颗粒。需要注意的是，大多数情况下体膨颗粒是指以部分中和的丙烯酸（AA）或丙烯酰胺（AM）为单体合成的。体膨颗粒在卫生、农林、油田、建筑等领域具有广泛的应用。

基于对酰胺键高温断裂水解的认识，采用耐高温四烯交联剂（四烯丙基氯化铵）和不耐高温酰胺类交联剂（N,N-亚甲基双丙烯酰胺）制备了具有耐高温和缓膨胀性质的温敏性体膨颗粒。本实验中合成的温敏性耐高温体膨颗粒作用机理如图 2.23 所示，低温下体膨颗粒两种交联物质同时发挥作用，交联密度大，体膨颗粒膨胀倍率低，高温下 N,N-亚甲基双丙烯酰胺形成的交联酰胺键水解，体膨颗粒交联密度下降，膨胀倍率增大。

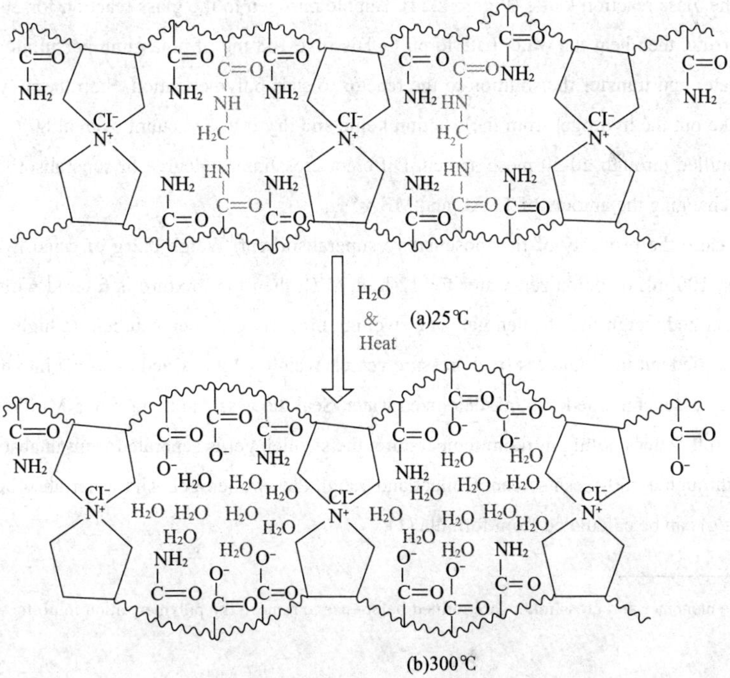

图 2.23　温敏性耐高温体膨颗粒作用机理示意图

3. Instruments and reagents

Instruments: Glass reaction kettle (500 mL); beaker (200 mL); stainless steel pressure vessel; water bath; magnetic stirrer; Muffle furnace; scanning electron microscopy (SEM); fourier transform infrared spectrum (FT-IR); thermal gravimetric analyzer (TGA); electronic balance; vacuum drying oven; thermometer; magneton.

Reagents: Acrylamide (AM), recrystallized with acetone/n-hexane before use; N, N-methylene bisacrylamide (BMA); tetraallylammonium chloride (TAAC); potassium persulfate (KPS); deionized water; high-purity nitrogen.

4. Experimental procedure

(1) Remove inhibitor from TAAC and MBA by alkaline alumina column ❶. Briefly, a glass burette with absorbent cotton on the tip is packed with basic aluminum oxide. The height of the self-made column is about 12 cm. 3 mL of MBA is passed through the column by using a rubber suction bulb. The purified MBA is collected for next polymerization. The refining method of TAAC is the same as that of MBA.

(2) Accurately weigh 28.4 g (0.4 moL) of acrylamide, 0.854 g of TAAC (4 mmoL) and 0.616 g of MBA (4 mmoL) into a 200 mL beaker, add 60 mL deionized water to dissolve, and then transfer the solution to the glass reaction kettle (Figure 2.24). Bubble nitrogen to the glass reaction for 30 min under magnetic stirring, then heat the water bath to 60 ℃. Dissolve 108 mg of potassium persulfate in 5 mL of deionized water and transfer the solution to the reactor to start polymerization. Stop the polymerization after 4 h. Take out the hydrogel from the reaction kettle and dry it in a vacuum oven at 80 ℃. The dried hrdrogel is milled through 20-50 mesh screen. Different crosslinking degree of superabsorbents can be obtained by changing the amount of TAAC and MBA ❷.

(3) Measure the property of thermosensitive superabsorbent. Weigh 0.2 g of dried hydrogel and incubate it in 100 mL of deionized water for 12 h at 25 ℃, then the mixture is filtered with 100 mesh screen. Collect and weigh the swollen gel. The swelling property of superabsorbent at high temperature needs to be carried out in a stainless steel pressure vessel. Weigh 0.2 g of dried hydrogel into the stainless steel pressure vessel contained 100 mL deionized water. Seal the vessel and put it in a Muffle furnace set to 300 ℃ for 6 h. After cooling to room temperature, the swollen gel is separated from unabsorbed water by filtering through a 100-mesh screen. Collect and weigh the swollen gel. The water absorbency of the sample (Q, g/g) can be calculated from formula (1).

---

❶ The monomer and crosslinkers are refined before use to remove the polymerization inhibitor contained in the reagent.

❷ The water absorbency of superabsorbents prepared by different crosslinking agent content and ratio can be tested in groups.

三、仪器和试剂

仪器：分体式玻璃反应釜（500 mL）；烧杯（200 mL）；陈化釜；水浴锅；磁力搅拌器；马弗炉；扫描电镜；傅里叶红外光谱；热重分析仪；电子天平；真空干燥箱；温度计；磁子。

试剂材料：丙烯酰胺（AM），使用前用丙酮/正己烷重结晶；N,N-亚甲基双丙烯酰胺（BMA）；四烯丙基氯化铵（TAAC）；过硫酸钾（KPS）；去离子水；高纯氮气。

四、实验步骤

（1）采用碱性三氧化二铝柱去除 N,N-亚甲基双丙烯酰胺和四烯丙基氯化铵中的阻聚剂❶。具体操作如下：取一根玻璃滴定管，在尖端塞上脱脂棉作为筛板，然后装入碱性三氧化二铝，高度 12 cm 左右。用吸管从试剂瓶中取 3 mL 的 N,N-亚甲基双丙烯酰胺加入自制的玻璃柱中，用吸耳球轻轻挤压，收集脱除阻聚剂的 N,N-亚甲基双丙烯酰胺后备用。四烯丙基氯化铵的精制方法同 N,N-亚甲基双丙烯酰胺。

（2）准确称取 28.4 g（0.4 mol）丙烯酰胺、0.854 g TAAC（4 mmol）、0.616 g MBA（4 mmol）置于 200 mL 烧杯中，再加入 60 mL 去离子水溶解后转入反应釜中（图 2.24）。在磁力搅拌状态下向反应釜中通氮气除氧 30 min，水浴锅升温至 60℃。称取 108 mg 过硫酸钾溶于 5 mL 去离子水中并加入反应釜，反应 4h 后将凝胶取出，放入 80℃ 真空烘箱中烘干，粉碎后过 20~50 目标准筛收集待用。改变 TAAC 和 MBA 的用量可以得到不同交联度的体膨颗粒❷。

（3）温敏性体膨颗粒性能测定。准确称取 0.2 g 左右的体膨颗粒均匀分散在 100 mL 的蒸馏水中，25℃下溶胀 12 h 后用 100 目标准筛过滤，然后收集并称重凝胶。体膨颗粒在高温下的膨胀性能需要在陈化釜中进行，称取 0.2 g 左右的体膨颗粒加入陈化釜中，加入 100 mL 去离子水密封后置于 300℃ 马弗炉中溶胀 6 h，冷却后用 100 目标准筛过滤，然后收集并称重凝胶，则体膨颗粒膨

---

❶ 单体和交联剂在使用前精制是为了除试剂中含有的阻聚剂。

❷ 可以分组考查不同交联剂含量和配比制备的体膨颗粒的膨胀倍率。

$$Q = \frac{m_2 - m_1}{m_1} \qquad (1)$$

Where, m_1 and m_2 are the mass of the dry sample and the swollen gel.

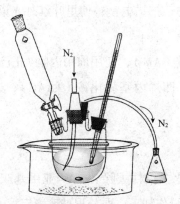

Figure 2.24 Schematic of polymerization device

(4) Spread the swollen hydrogel incubated at different temperatures on the mica sheet, and freeze the hydrogel with liquid nitrogen. Then, dry the samples by a vacuum freeze-drying machine. After 48 h of vacuum freeze-drying, pick up the samples and place them on a metal stub with double-sided conductive adhesive tape. Finally spray the samples with a thin gold film and observe their porous structure swollen at different temperature by SEM ❶.

(5) The thermal properties of thermosensitive high-temperature resistant superabsorbent can bemeasured by thermogravimetric analyzer in N_2 atmosphere. The heating rate is set at 10 ℃/min. The decomposition temperature of superabsorbents can be determined by observing their weight loss.

(6) By measuring the water absorbency (Q) of the sample at different temperatures with time, the swelling kinetic curves of the superabsorbents can be obtained.

(7) The chemical composition of the superabsorbents can be characterized by FTIR using KBr pellet method.

Questions

(1) What is the main way to adjust the swelling ratio of superabsorbents with temperature?
(2) Try to analyze the factors influencing the swelling ratio of the superabsorbents.

---

❶ The samples should be freeze dried before SEM observation but not vacuum dried.

胀倍率 Q（g/g）可以用式（1）计算。

$$Q = \frac{m_2 - m_1}{m_1} \tag{1}$$

式中，m_1 和 m_2 分别为干样品和凝胶样品的质量。

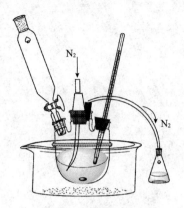

图 2.24　实验装置图

（4）将不同温度下吸水膨胀后的样品铺展在云母片上，用液氮快速冷却达平衡，然后再快速移入真空冷冻干燥机中，抽真空冷冻干燥 48h，取出喷金镀膜。通过扫描电镜观察体膨颗粒在不同温度下吸水膨胀后的网络结构❶。

（5）体膨颗粒的热性能采用热重分析仪在 N_2 氛围下进行测试，升温速率设置为 10℃/min。通过观察体膨颗粒的失重情况可以确定体膨颗粒的分解温度。

（6）通过测定体膨颗粒在不同温度下膨胀倍率（Q）随时间的变化可以得到体膨颗粒的膨胀动力学曲线。

（7）体膨颗粒的化学组成可以采用 KBr 压片法通过红外光谱仪表征。

思考题

（1）体膨颗粒的膨胀倍率随温度变化程度主要通过什么调节？

（2）试分析影响体膨颗粒膨胀倍率的影响因素。

---

❶ 通过扫描电镜观察不同温度下体膨颗粒的微观孔结构时必须采用冷冻干燥处理凝胶，不能采用真空干燥的方法。

第三章
Chapter 3

文献检索
Literature Search

3.1 SciFinder

SciFinder is the original database of Chemical Abstracts. It can search the related literature of chemistry, chemical engineering, medicine, biology, chemical substances and chemical reactions. It is an important tool for scientific research. If SciFinder is not installed on your computer, you can download SFS2007.exe client installation program from the library. It is also recommended to download viewerlite (software for viewing 3D models), and it is recommended to close all programs before installing the program. After confirming the successful networking, double-click the SciFinder Scholar icon on the desktop(Figure 3.1).

Figure 3.1

To start SciFinder Scholar double click the desktop icon:This opens the startup window of SciFinder(Figure 3.2):

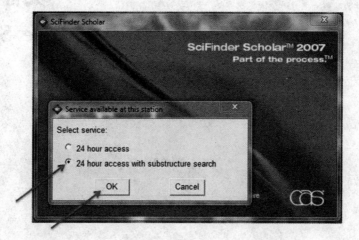

Figure 3.2

1. Choose ' 24 hour access with substructure search'

2. Click 'OK'

With a certain number of licenses of SciFinder, we can obtain all permissions according to the following operation requirements:

3. Press 'Cancel' and try again late

If you grant access to the database the following window opens:

3.1 SciFinder "数据库"

SciFinder 是美国化学学会（ACS）旗下的化学文摘服务社 CAS（Chemial Abstract Service）所出版的化学资料电子数据库学术版，可以检索化学、化工、医药、生物等相关文献，并可以进行化学物质、化学反应的检索，是科学研究的重要工具。如果电脑中没有安装 SciFinder，可以从学校图书馆下载 SFS2007.exe 客户端安装程序，同时建议下载 ViewerLite（查看 3D 模型的软件），安装程序前关闭所有程序。确认联网成功后双击桌面上的 SciFinder Scholar 图标，如图 3.1 所示。

图 3.1

这将会打开 SciFinder 的启动窗口，如图 3.2 所示。

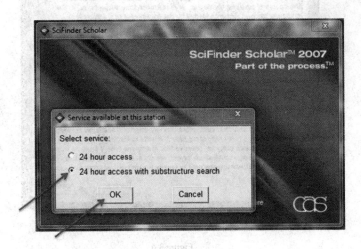

图 3.2

接下来的操作步骤如下。

（1）选择 "24 hour access with substructure search" 选项

（2）点击 "OK" 按钮

（3）点击 "Cancel" 按钮，稍后再试，如图 3.3 所示。出现如图 3.4 所示的

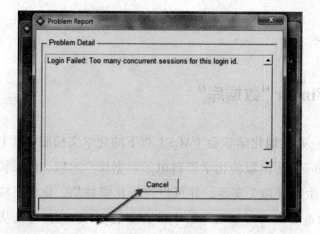

Figure 3.3

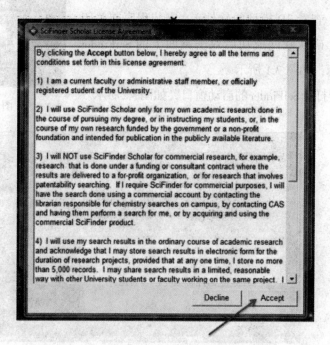

Figure 3.4

4. Accepting the SciFinder Scholar Agreements (click "Accept") opens the window "Message of the day".

5. Click 'OK' to open a new task:

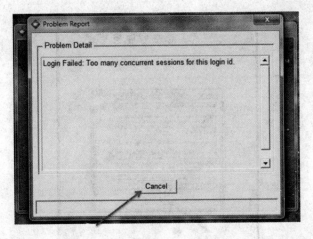

图 3.3

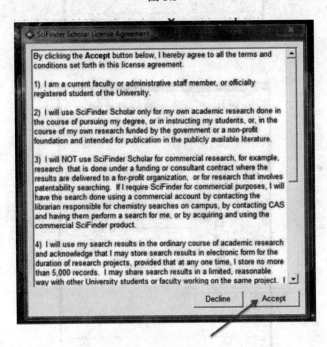

图 3.4

使用条款。

（4）接受 SciFinder 的使用条款（点击"Accept"按钮），会出现"Message of the day"窗口如图 3.5 所示。

（5）点击"OK"，打开新的对话框，如图 3.6 所示。

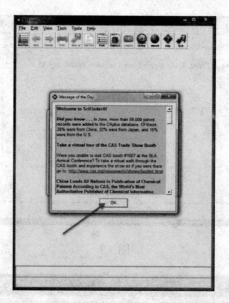

Figure 3.5

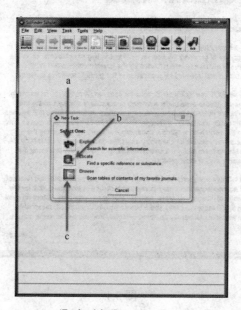

a. 'Explore'; b. 'Locate'; c. 'Browse'

Figure 3.6

3.1.1 SciFinder 'Explore'

Explore Literature

You can explore the literature by search for a specific topic, an author name or by company name/

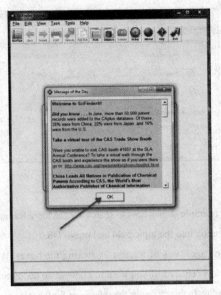

图 3.5

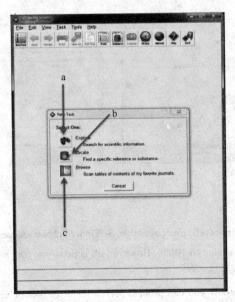

a—搜索；b—定位；c—浏览

图 3.6

3.1.1 SciFinder "搜索"

点击图 3.6 中的"搜索"即可进入搜索界面，如图 3.7 所示。

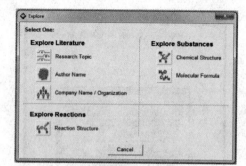

Figure 3.7

organization. The following example shows how to search for the topic 'radical polymerization':

Enter radical polymerization into the topic field and press 'OK':

You can select between candidates of interest by ticking the boxes. It is recommended to get the references containing the topic as entered. If you want to know more about general concepts to the entered topic, select the second possibility. In this example, the references for the topic as entered are selected:

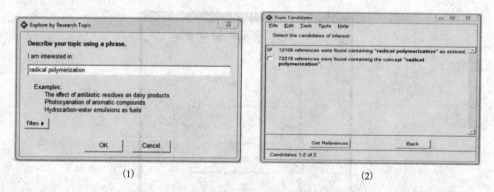

(1)   (2)

Figure 3.8

Due to the large number of references available, SciFinder Scholar reduces the amount of references for viewing, printing and saving to 10000. However, all actions you take will be executed to all the references available:

In the references window you can process search result further or view a reference you find interesting:

a. Some references are shown twice as they are accessible on two or more databases. By pressing 'Remove Duplicates' you can remove them.

b. By pressing 'Analyze/Refine' you can further decrease the number of references by further analyzing in terms of e.g. author name, publication year, chemicals of interest, etc.

第三章 文献检索
Chapter 3 Literature Search

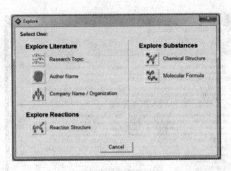

图 3.7

搜索文献。可以使用关键词，作者名或是机构的名称搜索文献。下面以关键词"自由基聚合"的搜索为例，简单介绍搜索的步骤。将自由基聚合输入，并点击"OK"键，如图 3.8 所示。

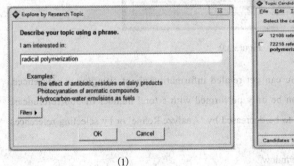
(1)

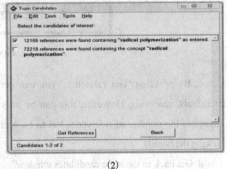
(2)

图 3.8

可以点击文献名前的按钮来选择感兴趣的文章。建议使用含有主题的关键词勾选图 3.8（2）中的第一个选择。如果想了解更多关于输入主题的一般概念，可以勾选图 3.8（2）中的第二个选择。以选择输入的主题搜索方法为例，点击"Get Reterences"即可进行搜索，如图 3.9 所示。

由于对现有文献的大量引用，SciFinder 把浏览、打印和保存的文献数量减少到了 1000 篇。具体操作将会影响到查阅到的文献数量。

在"引用"窗口中，可以处理搜索结果并进一步查看有价值的引用。

a. "Remove Duplicates" 当有些文献处在两个或是多个数据库时，它们在搜索时就会出现两次。通过按'Remove Duplicates'按钮可以除去重复的

Polymer Chemistry Experiments (Bilingual)
高分子化学实验

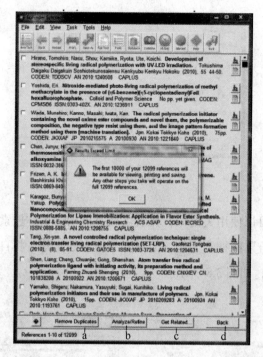

Figure 3.9

c. By pressing 'Get related...' you can get related information about citing/cited references, chemicals, reactions. However, this can be only performed with a total number of 500 references. Therefore the number of references has to be decreased by 'Analyze/Refine' or by selecting references (ticking the boxes).

d. Get back to the topic candidates window.

e. Get more details about the reference.

f. Search the ChemPort/SFX library of the FU Berlin for the selected literature.

Explore Substances

If you are searching for a specific substance you can search by drawing the chemical structure (e.g. if you do not know the correct name) or by entering the molecular formula.

Explore Reactions

If you want to search references/literature for a specific reaction (e.g. if you want to search for reaction conditions) you can enter the reaction here.

3.1.2 SciFinder 'Locate'

Locate Literature

If you are searching for specific literature, use this tool to search for publications and patents.

第三章 文献检索
Chapter 3 Literature Search | 163

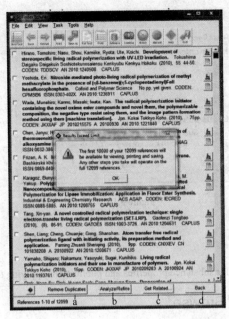

图 3.9

文献。

　　b.'Analyze/Refine'可以进一步对主题进行分析，如：作者、出版日期、感兴趣的化学物质等。这样就可以进一步减少搜索出的文献数量。

　　c."Get Related…"如果想得到更多关于引例文献、化学物质、反应等信息，可以按"Get Related…"查询，可能会得到大约 500 篇的文献，因此可以按'Analyze/Refine'按钮对文献进一步筛选。

　　d."Back"返回筛选窗口。

　　e."Back"得到关于文献的更多信息。

　　f."Back"搜索柏林自由大学 ChemPort/SFX 数据库，进一步筛选文献。

　　搜索物质。如果在不知道物质名称的情况下想查询一种物质，可以用搜索化学结构或分子式查找。

　　搜索反应。如果想搜索一个特定反应的文献，例如，在不能明确反应条件的情况下，可以直接搜索反应。

3.1.2　SciFinder "定位"

　　点击图 3.6 中的"定位"即可进入定位界面，如图 3.10 所示。

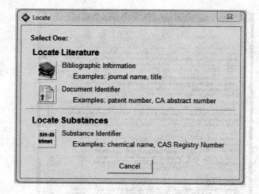

Figure 3.10

Locate Substances

Use this tool to search for substance properties and publications applying this substance.

3.1.3 SciFinder 'Browse'

Use this tool to browse the journal table of contents:

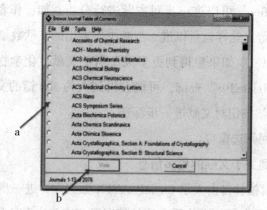

Figure 3.11

a. Select your journal of interest

b. Click 'View' to watch the current table of contents of your selected journal

第三章 文献检索
Chapter 3 Literature Search

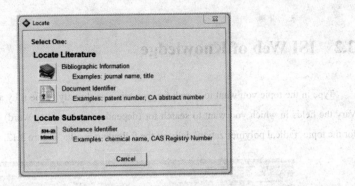

图 3.10

文献定位。如果想搜索特定的某篇文献，可以使用"SciFinder"搜索到出版物和专利。

物质定位。这个方法可以查询到物质特性并找到应用这种物质的文献。

3.1.3 SciFinder "浏览"

点击图 3.6 中的"浏览"即可进入浏览界面，如图 3.11 所示。用这种方法可以浏览期刊目录。

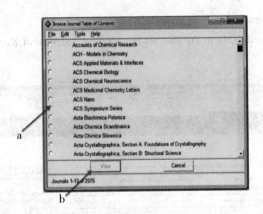

图 3.11

a. 选择感兴趣的期刊

b. 点击"View"按钮查看选择期刊的目录。

更详细的使用说明可在学校图书馆或网络上查阅培训课件"研究工具 SciFinder 介绍"。

3.2 ISI Web of Knowledge

Type in the topic you want to search for, you can also directly refine it by adding further key words. Vary the fields in which you want to search for (dependent on your key word). Click 'search' Searching for the topic 'radical polymerization' delivers the following results, Figure 3.12.

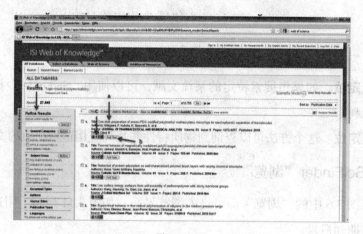

Figure 3.12

Now there are several possibilities to continue.

Figure 3.12(a) Go directly to the abstract provided by 'ISI Web of Knowledge' as showing in Figure.3.13.

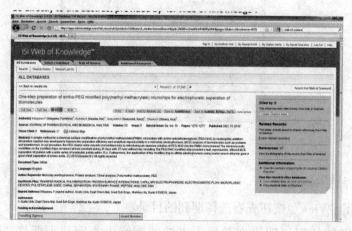

Figure 3.13

Figure 3.12(b) Directly go to the abstract/full text provided by a second database as showing in Figure 3.14.

3.2 ISI Web of Knowledge 平台

输入想搜索的标题,也可以增加关键词直接提炼,改变关键词的字段(取决于关键词),点击"search"按钮。例如,输入关键词"自由基聚合",得到如图 3.12 所示的结果。

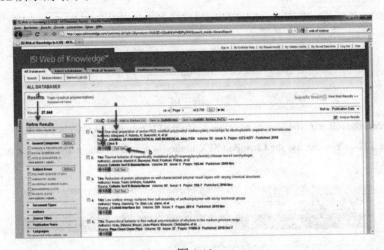

图 3.12

图 3.12(a)直接得到 ISI Web of Knowledge 平台提供的摘要如图 3.13 所示。

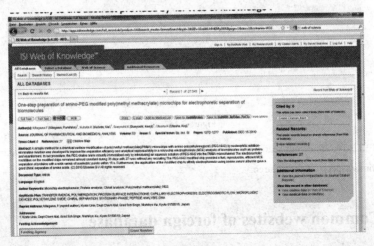

图 3.13

图 3.12(b)找到其他数据库提供的摘要或是全文如图 3.14 所示。

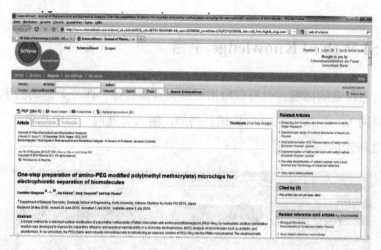

Figure 3.14

To access the full text, you have to be connected to the FU network (WLAN or LAN).

Figure 3.12(c) You can use the ChemPort/SFX library to search for alternative sources: as showing Figure 3.15:

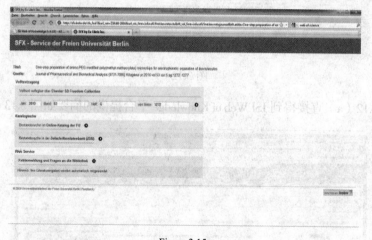

Figure 3.15

Figure 3.12(d) Refine your search by altering various aspects of your search. d. Refine your search by altering various aspects of your search.

3.3 Common websites of foreign database

You can also directly search for the required journals, books and patents through the website of foreign academic journals. The influential journal websites are as follows:

第三章 文献检索 | 169
Chapter 3 Literature Search

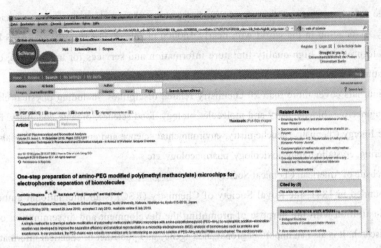

图 3.14

为得到全文内容，需要与 FU 网络（WLan or LAN）连接。

图 3.12（c）可以使用 ChemPort/SFX 数据库选择如图 3.15 所示。

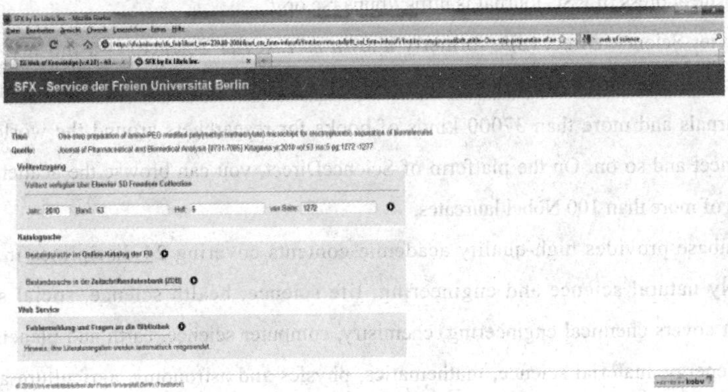

图 3.15

图 3.12（d）改变搜索的方向不断精炼搜索内容。

3.3 常用的外文期刊网站

通过外文期刊网站可以直接搜索到所需的期刊和文献，较有影响力的期刊网站如下。

(1) The American Chemical Society (ACS), founded in 1876, has become the largest science and Technology Association in the world with nearly 160000 members. For many years, ACS has been committed to providing high-quality literature information and services for universities, chemical research institutions and enterprises around the world.

ACS journals cover more than 20 chemistry related disciplines, including physical chemistry, chemical engineering, biochemistry and molecular biology, food science, organic chemistry, inorganic and Atomic Energy Chemistry, geochemistry, environmental science and engineering, materials science and engineering, crystallography, toxicology, pharmacology, etc.

The website of American Chemical Society (ACS) is https://pubs.acs.org/.

(2) Founded in 1841, the Royal Society of Chemistry (RSC) is an international authoritative academic institution with the purpose of promoting the development and dissemination of global chemical research. It is also an important propaganda agency and publisher of chemical information. The journals published by RSC are the core journals in the field of chemistry, most of which are included by SCI and MEDLINE. For example, Analyst, Chemical Society Reviews, Chemical Communications, Green Chemistry and so on are very famous journals in related fields. RSC electronic journal database includes 38 journals and 6 databases.

The website address of RSC journal is https://pubs.rsc.org/.

(3) Elsevier Science Direct (SD, formerly known as Elsevier Science) is a full-text database of scientific literature produced by Elsevier group in the Netherlands. It has provided more than 3800 peer-reviewed journals and more than 37000 kinds of books for researchers around the world, including Cell, The Lancet and so on. On the platform of ScienceDirect, you can browse the academic research achievements of more than 100 Nobel laureates.

The database provides high-quality academic contents covering 24 disciplines in four major fields, namely natural science and engineering, life science, health science, social science and humanities. It covers chemical engineering, chemistry, computer science, earth and planetary science, engineering, energy, material science, mathematics, physics and astronomy, agriculture and biology, biochemistry, genetics and molecular biology, environmental science, immunology and microbiology, nervous system science, medicine and stomatology, nursing and health, pharmacology, toxicology and pharmacology, veterinary science, arts and humanities, business, management and accounting, decision science, economics, econometrics and finance, psychology, social sciences, and interdisciplinary research.

Elsevier's website is https://www.sciencedirect.com/.

(4) Wiley company is one of the oldest and most famous academic publishers in the world. In 2010, Wiley online library was launched. There are more than 1600 peer-reviewed academic journals, 20000 e-books, 170 online reference books, 580 online reference books and 19 laboratory guides for biology, life science and biomedicine Protocols, 17 databases of chemistry, spectroscopy and evidence-based medicine (Cochrane Library).

（1）美国化学会（American Chemical Society，ACS）成立于1876年，现已成为世界上最大的科技学协会，拥有近16万会员。多年以来，ACS一直致力于为全球高校、化学研究机构和企业提供高品质文献资讯及服务，涵盖20多个与化学相关的学科，包括物理化学、化学工程、生物化学和分子生物学、食品科学、有机化学、无机与原子能化学、地球化学、环境科学与工程、材料科学与工程、晶体学、毒理学、药理学等诸多领域。网址：https://pubs.acs.org/

（2）英国皇家化学学会（Royal Society of Chemistry，RSC）成立于1841年，是以促进全球化学领域研究发展与传播为宗旨的国际权威学术机构，是化学信息的一个重要宣传机关和出版商。RSC出版的期刊是化学领域的核心期刊，大部分被SCI和MEDLINE收录，如Analyst、Chemical Society Reviews、Chemical Communications、Green Chemistry等都是相关领域中非常著名的期刊。RSC电子期刊数据库包括38种期刊和6个数据库。网址：https://pubs.rsc.org/。

（3）Elsevier ScienceDirect数据库（简称SD，曾用名Elsevier Science）是荷兰爱思唯尔（Elsevier）集团生产的科学文献全文数据库。为全球研究人员提供3800多种同行评审期刊和37000余种图书，包括全球影响力极高的Cell《细胞杂志》、The Lancet《柳叶刀杂志》等。在ScienceDirect平台上可以浏览100余位诺贝尔奖获得者的学术研究成果。数据库提供覆盖自然科学与工程、生命科学、健康科学、社会科学与人文科学4大领域24个学科的优质学术内容，涉及化学工程、化学、计算机科学、地球与行星学、工程、能源、材料科学、数学、物理学与天文学、农业与生物学、生物化学、遗传学和分子生物学、环境科学、免疫学和微生物学、神经系统科学、医学与口腔学、护理与健康、药理学、毒理学和药物学、兽医科学、艺术与人文科学、商业、管理和财会、决策科学、经济学、计量经济学和金融、心理学、社会科学，以及交叉研究领域。网址：https://www.sciencedirect.com/。

（4）Wiley公司是全球历史最悠久、最知名的学术出版商之一。2010年推出在线资源平台Wiley Online Library，该平台上有1600多种经同行评审的学术期刊，20000多种电子图书，170多种在线参考工具书，580多种在线参考书，19种生物学、生命科学和生物医学的实验室指南（Current Protocols），17种化学、光谱和循证医学数据库（Cochrane Library）。

Wiley's journals cover a wide range of disciplines, including chemistry, polymer and materials science, physics, engineering, medicine, nursing, stomatology, life sciences, business, economics, history, political science, sociology, art, anthropology, and many other important interdisciplinary journals.

The website of Wiley online library is http://onlinelibrary.wiley.com/.

(5) Springer was founded in Berlin, Germany in 1842. It is the world's largest publisher of STM (Science, technology and medicine) books and the second largest publisher of STM journals. It publishes more than 8400 kinds of scientific books and more than 2200 leading scientific journals every year.

Springer publishes more than 2200 journals each year, covering 11 disciplines, including natural science, technology, engineering, medicine, law, behavioral science, economics, biology and medicine. More than 60% of the journals published by Springer are included in SCI and SSCI, and some journals have high ranking in related disciplines.

The website of Springer is http://link.springer.com/.

(6) Nature is the earliest international science and technology journal in the world. Since it was founded in 1869, Nature has consistently reported and commented on the most important breakthroughs in the global science and technology field. The electronic journals provide by Nature.com cover science, technology, biotechnology, chemistry, gene and evolution, immunology, pharmacy, medicine, clinical medicine, malignant tumor, dentistry, molecular cell biology, neuroscience, physical science, etc. The popular journals of nature full-text database include Nature, Nature Chemistry, Nature Communication, Nature Materials, etc.

The website of Nature is https://www.nature.com/.

(7) Science was founded by Edison in 1880 and published by the American Association for the advancement of Science (AAAS) in 1900. It is a comprehensive scientific journal with high reputation in the international academic community. Science online is the online database of science, involving life sciences and medicine, basic natural sciences, engineering, and some humanities and social sciences.

The website of Science Online is http://www.sciencemag.org/

Wiley 的期刊学科范围广，包括化学、高分子与材料科学、物理学、工程学、医学、护理学、口腔医学、生命科学、商业、经济、历史学、政治学、社会学、艺术类、人类学等学科，以及很多其他重要的跨学科领域出版的期刊。网址：http://onlinelibrary.wiley.com/。

（5）施普林格（Springer）公司于 1842 年在德国柏林创立，是全球第一大 STM（科学、技术和医学）图书出版商和第二大 STM 期刊出版商，每年出版 8400 余种科技图书和 2200 余种领先的科技期刊。涵盖自然科学、技术、工程、医学、法律、行为科学、经济学、生物学和医学等 11 个学科。Springer 出版的期刊 60% 以上被 SCI 和 SSCI 收录，一些期刊在相关学科拥有较高的排名。网址：http://link.springer.com/。

（6）英国著名杂志《Nature》是世界上最早的国际性科技期刊，1869 年创刊以来，始终如一地报道和评论全球科技领域里最重要的突破。Nature.com 平台提供的电子期刊，主题涵盖科学、技术、生物技术、化学、基因与进化、免疫、药学、医学、临床医学、恶性肿瘤、牙科、分子细胞生物、神经科学、物理科学等。大家熟悉的 Nature 全文数据库的期刊包括 Nature、Nature Chemistry、Nature Communication、Nature Materials 等。网址：https://www.nature.com/。

（7）美国《科学》周刊（Science）由爱迪生于 1880 年创建，1900 年开始由美国科学促进会（American Association for the Advancement of Science，简称 AAAS）负责出版，是在国际学术界享有盛誉的综合性科学期刊。Science Online 是《科学》杂志的网络数据库，涉及生命科学及医学、基础自然科学、工程学，以及部分人文社会科学。网址：http://www.sciencemag.org/。

附录Ⅰ 高分子相关的主要学术期刊

刊名	CSCD 大类	大类分区	3 年平均 IF（2016—2018）
Nature Materials	工程技术	1	39.286
Materials Today	工程技术	1	23.535
Advanced Materials	工程技术	1	22.517
ACS Nano	工程技术	1	13.851
Advanced Functional Materials	工程技术	1	13.69
Materials Horizons	工程技术	1	12.748
Chemistry of Materials	工程技术	1	9.838
Small	工程技术	1	9.699
Biomaterials	工程技术	1	9.160
ACS Applied Materials & Interface	工程技术	1	8.019
Applied Materials Today	工程技术	1	8.013
Polymer Reviews	工程技术	1	6.638
Acta Biomaterialia	工程技术	1	6.447
Acta Materialia	工程技术	1	6.210
Materials Research Letters	工程技术	1	6.125
Journal of Materials Chemistry C	工程技术	1	5.958
Applied Surface Science	工程技术	2	4.327
Reactive and Functional Polymers	工程技术	2	3.067
Chemical Reviews	化学	1	51.614
Chemical Society Reviews	化学	1	39.748
Progress in Polymer Science	化学	1	24.943
Journal of the American Chemical Society	化学	1	14.303
Angewandte Chemie-International Edition	化学	1	12.118
Chemical Science	化学	1	9.096
Chemical Communications	化学	1	6.258
Analytical Chemistry	化学	1	6.237
Science China-Chemistry	化学	1	4.888

续表

刊名	CSCD 大类	大类分区	3 年平均 IF（2016—2018）
Advances in Colloid and Interface Science	化学	2	7.604
Macromolecules	化学	2	5.915
Biomacromolecules	化学	2	5.550
Carbohydrate Polymers	化学	2	5.338
Journal of Colloid and Interface Science	化学	2	5.228
Polymer Chemistry	化学	2	5.021
Langmuir	化学	2	3.768
Soft Matter	化学	2	3.666
Polymers	化学	2	3.154
Plasma Process and Polymers	物理	2	2.906
Polymer	化学	3	3.646
European Polymer Journal	化学	3	3.631
Journal of Polymer Science Part A	化学	3	2.710
Polymer International	化学	3	2.285
Polymer Composite	化学	3	2.178
高分子学报	化学	3	0.831
功能高分子学报	化学		
高分子材料科学与工程	化学		
高分子通报	化学		
离子交换与吸附	化学		
胶体与聚合物	化学		

附录Ⅱ 常见聚合物的中英文对照及缩写

中文名称	英文名称	英文缩写
丙烯腈-丁二烯-苯乙烯共聚物	Acrylonitrile Butadiene Styrene	ABS
醇酸树脂	Alkyd resin	AR
丁二烯橡胶	Butadiene rubber	BR
醋酸纤维素	Cellulose acetate	CA
羧甲基纤维素	Carboxymethyl cellulose	CMC
硝酸纤维素	Cellulose nitrate	CN
氯化聚乙烯	Chlorinated polyethylene	CPE
顺式-聚异戊二烯	*cis*-polyisoprene	CPI
氯化聚氯乙烯	Chlorinated polyvinylchloride	CPVC
氯丁橡胶	Chloroprene rubber	CR
环氧树脂	Epoxy resin	ER
高密度聚乙烯	High density polyethylene	HDPE
丁基橡胶	Butyl rubber	IIR
低密度聚乙烯	Low density polyethylene	LDPE
甲基纤维素	Methyl cellulose	MC
天然橡胶	Natural rubber	NR
聚酰胺（尼龙）	Polyamide（Nylon）	PA
聚己二酰己二胺（尼龙-66）	Poly（hexamethylene adipamide）（Nylon-66）	PA-66
聚己内酰胺（尼龙-6）	Polycaprolactam（Nylon-6）	PA-6
聚丙烯酸	Poly（acrylic acid）	PAA
聚丙烯酰胺	Polyacrylamide	PAAM
聚丙烯腈	Polyacrynitrile	PAN
聚丁二烯	Polybutadiene	PB
聚丙烯酸丁酯	Poly（butyl acrylate）	PBA
聚甲基丙烯酸正丁酯	Poly（butyl methacrylate）	PBMA

续表

中文名称	英文名称	英文缩写
聚对苯二甲酸丁二醇酯	Poly (butylene terephthalate)	PBT
聚碳酸酯	Polycarbonate	PC
聚甲基丙烯酸-N,N-二甲胺乙酯	Poly [2-(N,N-dimethyl amino) ethyl methacrylate]	PDMAEMA
聚乙烯	Polyethylene	PE
聚丙烯酸乙酯	Poly (ethyl acrylate)	PEA
聚醚醚酮	Poly (ether ether ketone)	PEEK
聚乙二醇	Poly (ethylene glycol)	PEG
聚甲基丙烯酸乙酯	Poly (ethyl methacrylate)	PEMA
聚环氧乙烷	Poly (ethylene oxide)	PEO
聚对苯二甲酸乙二醇酯	Poly (ethylene terephthalate)	PET
聚丙烯酸羟乙酯	Poly (hydroxyethyl acrylate)	PHEA
聚甲基丙烯酸羟乙酯	Poly (hydroxyethyl methacrylate)	PHEMA
酚醛树脂	Phenol-formaldehyde	PF
聚异戊二烯	Polyisoprene	PI
聚异丁烯	Polyisobutylene	PIB
聚甲基丙烯酸异丁酯	Poly (isobutyl methacrylate)	PIBMA
聚丙烯酸缩水甘油酯	Poly (glycidyl acrylate)	PGA
聚丙烯酸甲酯	Poly (methyl acrylate)	PMA
聚甲基丙烯酸甲酯	Poly (methyl methacrylate)	PMMA
聚 N-异丙基丙烯酰胺	Poly (N-isopropylacrylamide)	PNIPAM
聚烯烃	Polyolefin	PO
聚甲醛	Polyoxymethylene	POM
聚丙烯	Polypropylene	PP
聚环氧丙烷	Poly (propylene oxide)	PPO
聚苯硫醚	Poly (phenylene sulphide)	PPS
聚苯乙烯	Polystyrene	PS
聚四氟乙烯	Polytetrafluoroethylene	PTFE
聚氨酯	Polyurethane resin	PU
聚乙烯醇	Poly (vinyl alcohol)	PVA
聚醋酸乙烯酯	Poly (vinyl acetate)	PVAc

续表

中文名称	英文名称	英文缩写
聚氯乙烯	Poly（vinyl chloride）	PVC
聚偏氯乙烯	Poly（vinylidene chloride）	PVDC
聚偏氟乙烯	Poly（vinylidene fluoride）	PVDF
聚氟乙烯	Poly（vinyl fluoride）	PVF
聚乙烯基咔唑	Poly（vinyl carbazole）	PVK
聚乙烯吡咯烷酮	Poly（vinyl pyrrolidone）	PVP
聚2-乙烯基吡啶	Poly（2-vinyl pyridine）	P2VP
丁苯橡胶	Styrene-butadiene rubber	SBR
苯乙烯-丁二烯-苯乙烯共聚物	Poly（styrene-b-butadiene-b-styrene）	SBS
脲醛树脂	Urea-formaldehyde	UF
超高分子量聚乙烯	Ultrahigh molecular weight polyethylene	UHMWPE
不饱和树脂	Unsaturated polyesters	UP

附录Ⅲ 一些常见的高分子溶剂和沉淀剂

名称	溶剂	沉淀剂
聚丁二烯（PB）	脂肪烃、芳香烃、卤代烃、四氢呋喃、高级酮和酯	水、醇、丙酮等
聚乙烯（PE）	甲苯、二甲苯、十氢化萘、四氢化萘	醇、丙酮、邻苯二甲酸甲酯
聚丙烯（PP）	环己烷、二甲苯、十氢化萘、四氢化萘	醇、丙酮、邻苯二甲酸甲酯
聚异丁烯（PIB）	烃、氯代烃、四氢呋喃、高级脂肪醇和酯、二硫化碳	低级酮、低级醇、低级酯
聚氯乙烯（PVC）	丙酮、环己酮、四氢呋喃	醇、己烷、氯乙烷、水
聚四氟乙烯（PTFE）	全氟煤油（350℃）	大多数溶剂
聚丙烯酸（PAA）	乙醇、二甲基甲酰胺、水、稀碱溶液、二氧六环/水（8/2）	脂肪烃、芳香烃、丙酮、二氧六环
聚丙烯酸甲酯（PMA）	丙酮、丁酮、苯、甲苯、四氢呋喃	水、甲醇、乙醇、乙醚
聚丙烯酸乙酯（PEA）	丙酮、丁酮、苯、甲苯、四氢呋喃、丁醇	脂肪醇（C≥5）、环己醇
聚丙烯酸丁酯（PBA）	丙酮、丁酮、苯、甲苯、四氢呋喃、丁醇	甲醇、乙醇、乙酸乙酯
聚甲基丙烯酸（PMAA）	乙醇、水、稀碱溶液、盐酸［0.02（mol/L），30℃］	脂肪烃、芳香烃、丙酮、羧酸、酯
聚甲基丙烯酸甲酯（PMMA）	丙酮、丁酮、苯、甲苯、四氢呋喃、氯仿、乙酸乙酯	甲醇、石油醚、己烷、环己烷、水
聚甲基丙烯酸乙酯（PEMA）	丙酮、丁酮、苯、甲苯、四氢呋喃、乙醇（热）	异丙醚
聚甲基丙烯酸异丁酯（PIBMA）	丙酮、乙醚、汽油、四氯化碳、乙醇（热）	甲醇、乙醇（冷）
聚甲基丙烯酸异丁酯（PBMA）	丙酮、丁酮、苯、甲苯、四氢呋喃、己烷、正己烷	甲醇、乙醇（冷）
聚醋酸乙烯酯（PVAc）	丙酮、苯、甲苯、四氢呋喃、氯仿、二氧六环	无水乙醇、己烷、环己烷
聚乙烯醇（PVA）	水、乙二醇（热）、丙三醇（热）	烃、卤代烃、丙酮、丙醇

续表

名称	溶剂	沉淀剂
聚乙烯醇缩甲醛（PVFM）	甲苯、氯仿、2-氯乙醇、苯甲醇、四氢呋喃	脂肪烃、甲醇、乙醇、水
聚丙烯酰胺（PAAM）	水	醇类、四氢呋喃、乙醚
聚甲基丙烯酰胺（PMAAm）	水、甲醇、丙酮	酯类、乙醚、烃类
聚N-异丙基丙烯酰胺（PNIPAM）	水（<32℃）、苯、四氢呋喃	水（>32℃）、正己烷
聚甲基乙烯基醚（PMVE）	苯、氯代烃、正丁醇、丁酮	庚烷、水
聚丁基乙烯基醚（PBVE）	苯、氯代烃、正丁醇、丁酮、乙醚、正庚烷	乙醇
聚丙烯腈（PAN）	N,N-二甲基甲酰胺、乙酸酐	烃、卤代烃、酮、醇
聚苯乙烯（PS）	苯、甲苯、氯仿、环己烷、四氢呋喃、苯乙烯	醇、酚、己烷、丙酮
聚2-乙烯基吡啶（P2VP）	氯仿、乙醇、苯、四氢呋喃、二氧六环、吡啶、丙酮	甲苯、四氯化碳
聚4-乙烯基吡啶（P4VP）	甲醇、苯、环己酮、四氢呋喃、吡啶、丙酮/水（1:1）	石油醚、乙醚、丙酮、乙酸乙酯、水
聚乙烯基吡咯烷酮（PVP）	氯仿、甲醇、乙醇	烃类、四氯化碳、乙醚、丙酮、乙酸乙酯
聚氨酯（PU）	苯酚、甲酸、N,N-二甲基甲酰胺	饱和烃、醇、醚
聚对苯二甲酸乙二醇酯（PET）	苯酚、硝基苯（热）、浓硫酸	醇、酮、醚、烃、卤代烃
聚2,6-二甲基苯醚（PMPO）	苯、甲苯、氯仿、二氯甲烷、四氢呋喃	甲醇、乙醇

附录Ⅳ 常见引发剂的提纯方法

引发剂	提纯方法	使用温度/℃
过氧化苯甲酰（BPO）	将 BPO 粗品溶于三氯甲烷，再加等体积的甲醇或石油醚使 BPO 结晶析出。也可采用丙酮加两倍体积的蒸馏水重结晶。示例：将 5g BPO 室温下溶于 20mL 的三氯甲烷，过滤除去不溶性杂质，滤液滴入等体积的甲醇中结晶，过滤，晶体用冷甲醇洗涤，室温下真空干燥，冰箱中避光保存	60~100
偶氮二异丁腈（AIBN）	可用丙酮、三氯甲烷或甲醇重结晶。示例：在 100mL 锥形瓶中加入 50mL 95% 的乙醇接近沸腾，迅速加入 5g AIBN 溶解，趁热过滤，滤液冷却结晶，再次过滤得到晶体后真空干燥，避光贮存于冰箱中	50~90
偶氮二异庚腈（ABVN）	可用甲醇或乙醇重结晶。示例：在 200mL 圆底烧瓶中加入 100mL 95% 乙醇，水浴 80℃，迅速加入 10g ABVN 溶解，趁热过滤，滤液冷却结晶，再次过滤得到晶体后真空干燥，避光贮存于冰箱中	20~80
过氧化二异丙苯（DCP）	95% 乙醇溶解后，活性炭脱色，冷却至 0℃ 以下结晶，过滤后得到晶体并真空干燥，避光贮存于冰箱中	110~150
过硫酸钾（KPS）/过硫酸铵（APS）	过硫酸盐作为水溶性引发剂，其主要杂质是硫酸氢盐和硫酸盐，采用去离子水重结晶。示例：在 100mL 圆底烧瓶中加入 50mL 去离子水，水浴 40 ℃，加入 5g KPS 溶解，趁热过滤，滤液冷却结晶，再次过滤得到晶体后真空干燥，避光贮存于冰箱中	50~90

参考文献

[1] 梁晖,卢江. 高分子化学实验[M]. 北京:化学工业出版社,2004.

[2] 孙汉文,王丽梅,董建. 高分子化学实验[M]. 北京:化学工业出版社,2012.

[3] 何卫东,金邦坤,郭丽萍. 高分子化学实验[M]. 合肥:中国科学技术大学出版社,2012

[4] CE 席尔奈希特,等. 聚合过程[M]. 唐士培,等译. 北京:化学工业出版社,1984.

[5] 邓云祥,刘振兴,冯开才. 高分子化学、物理和应用基础[M]. 北京:高等教育出版社,1997.

[6] 潘祖仁. 高分子化学[M]. 5 版. 北京:化学工业出版社,2014.

[7] 上海珊瑚化工厂. 有机玻璃[M]. 上海:上海人民出版社,1975.

[8] 赵德仁,张慰盛. 高聚物合成工艺学[M]. 5 版. 北京:化学工业出版社. 2014.

[9] Odian G. Principles of Polymerization (4th Edition)[M]. New Jersey:John Wiley & Sons,Inc,2004.

[10] CE 席尔奈希特,等. 高分子方法[M]. 朱秀昌,等译. 北京:科学出版社,1984.

[11] 黄美玉,等. 超高吸水性聚丙烯酸钠的制备[J],高分子通讯,1984,2:129-132.

[12] 赵德仁. 高聚物合成工艺学[M]. 北京:化学工业出版社,1981.

[13] 北京大学化学系高分子教研室. 高分子实验与专论[M]. 北京:北京大学出版社,1990.

[14] E.L. 麦卡弗里著. 蒋硕建译. 高分子化学实验室制备[M]. 北京:科学出版社,1981.

[15] 方禹声. 聚氨酯泡沫塑料[M]. 2 版. 北京:化学工业出版社,1994.

[16] 应圣康,郭少华. 离子型聚合[M]. 北京:化学工业出版社,1985.

[17] 复旦大学高分子科学系高分子科学研究所. 高分子实验技术[M]. 上海:复旦大学出版社,1996.

[18] Keddie D J,Moad G,Rizzardo E,etal. RAFT agent design and synthesis[J]. Macromolecules,2012,45:5321-5342.

[19] Moad G,Rizzardo E,Thang S H. Living radical polymerization by the RAFT process —a second update[J]. Aust. J. Chem.,2009,62:1402-1472.

[20] Braun D, Cherdron H, Rehahn M, Ritter H, Voit B, Polymer Synthesis: Theory and Practice: Fundamentals, Methods, Experiments (4th Edition) [M], Springer, 2004.

[21] Smith W F. Principles of materials science and engineering [M]. United States, 1986.

[22] Ram Arie. Fundamentals of Polymer Engineering [M]. New York: Plenum Press, 1997.

[23] Seymour R B. Polymers for Engineering Applications [M]. ASM International, 1987.

[24] Allcock H R, Lampe R W; Mark J E. Contemporary Polymer Chemistry (3rd Edition) [M]. Pearson Education: Upper Saddle River, NJ, 2003.

[25] Zhang X, Tang Z, Tian D, et al. A self-healing flexible transparent conductor made of copper nanowires and polyurethane [J], Materials Research Bulletin, 2017. 90: 175-181.

[26] 汤周. 基于铜纳米线的导电复合材料的制备及研究 [D]. 2017, 中国石油大学（华东）.

[27] 肖力光, 张猛. 无皂乳液聚合理论及其研究进展 [J]. 吉林建筑大学学报, 2015. 32 (5): 5-8.

[28] 朱雯, 黄芳婷, 董观秀, 张明. 无皂乳液聚合法制备单分散聚苯乙烯微球 [J]. 功能材料, 2012. 43 (6): 775-778.

[29] Goodall A. R., Wilkinson M.C., Hearn J. Mechanism of emulsion polymerization of styrene in soap-free systems [J]. Journal of Polymer Science, 1977.

[30] 张晓云, 吴伟. 自由基水溶液聚合的改进 [J]. 高分子通报, 2015. 5: 101-103.

[31] 徐燕莉. 表面活性剂的功能 [M]. 北京: 化学工业出版社, 2000.

[32] Zhou W Q, Gu T Y, Su Z G, Ma G H. Synthesis of macroporous poly (styrene-divinyl benzene) microspheres by surfactant reverse micelles swelling method [J]. Polymer, 2007. 48: 1981-1988.

[33] 王向鹏. 耐高温体膨颗粒的制备及其在稠油热采防汽窜中的应用研究 [D]. 中国石油大学（华东）, 2015.

[34] Zhang X Y, Wang X P, Li L, et al. A novel polyacrylamide-based superabsorbent with temperature switch for steam breakthrough blockage [J]. Journal of Applied Polymer Science, 2015. 42067: 1-7.

[35] Zhang X Y, Wang X P, Li L, et al. Preparation and swelling behaviors of a high temperature resistant superabsorbent using tetraallylammonium chloride as crosslinking agent [J]. Reactive & Functional Polymers, 2015. 87: 15-21.